Electronic Displays

Library of Congress Cataloging in Publication Data
Bylander, E G
    Electronic displays.

    (Texas Instruments electronic series)
    Includes index.
    1. Information display systems. I. Title.
TK7882.I6B94   621.3815'42   78-31849
ISBN 0-07-009510-8

*The editors for this book were Tyler G. Hicks and Joseph Williams,
and the production supervisor was Sally Fliess. It was
set in Times Roman by University Graphics, Inc.*

Printed and bound by Halliday Lithographic, Inc.

Texas Instruments reserves the rights to make changes at any time in order to
improve design and supply the best product possible. Information contained
in this publication is believed to be accurate and reliable. However, responsi-
bility is assumed neither for its use nor for any infringement of patent or
rights of others which may result from its use. No license is granted by
implication or otherwise under any patent or patent right of Texas Instruments or others.

Contents

Preface ix

Chapter 1. Introduction to Electronic Displays 1

    1.1 Introduction 1
    1.2 Survey of Display Types 1
    1.3 Applications 3
    1.4 The Electronic Display Symbols 3

Chapter 2. Electronic Display Fundamentals 11

    2.1 Specifying Displays 11
        Introduction 11
        Major Design Factors 13
    2.2 Comparison Methods 13
    2.3 Addressing Methods 14
    2.4 Connectors and Mounting 18
    2.5 General Test Methods 19
        Test Philosophy 19
        Test Parameters 21
    APPENDIX 2.1: Multiplexing of Matrix Displays 24

Chapter 3. Display Human Factors 29

    3.1 Introduction to Display Human Factors 29
    3.2 Visibility 29
        Definitions of Contrast 30
        Photometry 33
        Contrast Enhancement 39
        Color Contrast 44
        Human Factors: Minimum Contrast 48
        Summary 50
    3.3 Legibility 50
        Character Size and Proportion 51
        Sharpness and Blur 51
        Summary 52
    3.4 Temporal Factors 52

APPENDIX 3.1: Photometric Units: A Glossary of Terms 55
APPENDIX 3.2: Conversion of Candelas to Footlamberts 57

Chapter 4. Gas Discharge Displays 59

4.1 Introduction 59
4.2 Gas Discharge Fundamentals 59
    Gas Discharge Geometry 59
    Current-Voltage Relationships of the Discharge 61
    Applications to the Display 63
    Fill Gases 65
    Light Emission 67
    AC Plasma Display 68
4.3 The Gas Discharge Display 70
    Gas Discharge Display Applications 72
    Construction Techniques 72
    Forming 74
    Failure Mechanisms and Temperature Considerations 74
    Special Provisions and Requirements 77
4.4 DC Drive Circuits 78
    Anode Drivers 78
    Segment Driver Circuit 78
4.5 AC Drive Circuits 83
APPENDIX 4.1: Analysis of Multiplexed Multiple-Segment Tube 88
APPENDIX 4.2: Parameter Estimation 95

Chapter 5. The Visible Light-Emitting Diode Display 99

5.1 Introduction 99
5.2 Principles of Operation 99
    VLED Material 100
    Diode Operation 102
    Color 108
    VLED Brightness 108
5.3 The VLED Display 110
    Display Configurations 110
    Fabrication 113
    Reliability and Environmental Considerations 114
    Special Considerations 115
5.4 Drive Circuit Requirements 115
    General Driving Requirements 115
    Multiplexed Displays 117
APPENDIX 5.1: Microprocessor Interfacing 121

Chapter 6. Vacuum Fluorescent Display 123

6.1 Introduction 123
6.2 Principles of Operation 123
    Tube Display Configuration 123
    Tube Operation 125
    Phosphor Principles 126
6.3 The VF Display 127
    Fabrication Methods 127
    Reliability and Environmental Considerations 129
    Special Considerations 129

6.4 Drive Circuits 131
APPENDIX 6.1: Calculation of Tube Currents 131

Chapter 7. Liquid Crystal Displays 137

7.1 Introduction 137
7.2 Liquid Crystal Principles 137
    LCD Materials 137
    Optics 139
    Failure Modes 141
7.3 Liquid Crystal Display 142
    Operating Parameters 143
    Special Considerations 147
7.4 Multiplexing Principles 147
    LC MUX Properties 148
APPENDIX 7.1: An Optimum Set of Multiplexing Voltages 152
APPENDIX 7.2: Pulse Design 153

Chapter 8. The Incandescent Display and the Cathode-Ray-Tube Display ... 157

8.1 The Incandescent Display 157
    Introduction 157
    Operation 157
    Construction 158
    Reliability and Environmental Considerations 159
    Multiplex Operation 159
8.2 The Cathode-Ray-Tube Display 160
    Introduction 160
    Cathode-Ray-Tube Considerations 160
    Notes on Character Display 160
APPENDIX 8.1: Incandescent Display Parameter Estimation 164

Index 167

Preface

Digital electronics has become pervasive as a result of its extremely low cost. Displays for digital systems allow direct indication and reading of numbers, letters, and symbols. They can present more data in less space than the obsolescent analog meters.

This book about the application of digital displays is for the practicing engineer. It is written to answer recurring questions that arise, and its purpose is to bridge the interface between the display designer and manufacturer in the middle and between the prospective user (or user engineer) on one side and the circuit designer on the other. To the extent that this book is an aid to this task, it will be successful in its purpose.

The book first covers general display considerations, such as font, legibility, size, and comparison methods. It then explores several display types—such as the gas discharge, visible light-emitting diode, vacuum fluorescent, and liquid crystal displays—at some length. Mentioned briefly are the incandescent and cathode-ray-tube displays.

The dc gas discharge and liquid crystal displays are difficult to drive; their operating conditions are explored in somewhat greater detail than those for the more easily driven displays. An attempt has also been made to give the practicing engineer a feel for the important approaches to specifying the user interface.

Acknowledgments: Years of interaction with internal and external Texas Instruments customers have led to the viewpoints expressed in this book; the continuing help of these customers is acknowledged. The optoelectronics marketing department in the persons of Michael S. Bender and Carroll E. Smith supplied advice and references.

The continuing support of Carroll E. Nelson, John W. Vance, and Harold L. Woody is also appreciated. The typing was carefully done by Margaret Grigg, who was of great assistance. Finally, any errors are solely the author's.

E. G. Bylander

Electronic Displays

1

Introduction to Electronic Displays

1.1 INTRODUCTION

The digital era has led to the obsolescence of the analog art. Historical analog applications of sensing and control are now digital. While much design emphasis is concentrated on the microprocessor [the arithmetic logic unit (ALU) or central processor unit (CPU)], its peripheral input and output requirements must be met as well. That is, digital means must be provided to communicate with the digital system sensors or analog sensors; and analog-to-digital (A–D) converters and digital or alphanumeric displays, printers, and terminals are required. To aid digital design, the selection criteria and application techniques for the electronic display portion of the digital system are described here.

Chapter 1 describes display choices, typical applications, and fonts. Chapter 2 is concerned with general display principles and applications, and Chapter 3 deals with viewing or human factors considerations. Subsequent chapters take up the major display classes individually.

1.2 SURVEY OF DISPLAY TYPES

Displays may be classified by several schemes. A display family tree (Fig. 1.1) classifies displays by segmental, or dot, matrix; by number and size of characters; and by emissive or passive mechanism. A number of display types will not be further discussed, including image, analog segmental, and electromechanical. Popular emissive mechanisms used for displays are summarized in Table 1.1; absorptive processes are grouped in Table 1.2. From these classifications one can obtain, for example, the cathode-ray-tube character generator combination for alphanumeric applications, the flat-panel gas discharge display, the liquid crystal display (LCD) and the visible light-emitting diode (VLED) displays, and the dot-matrix plasma panel display. In Table 1.3 some commercial realizations of such displays are listed. These displays will be considered later in individual chapters. Additionally, a good display bibliography is given as a reference.[4]

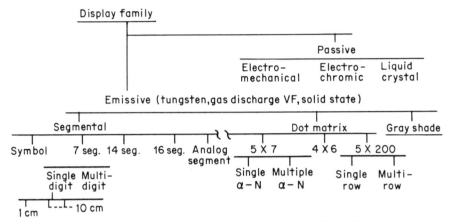

Fig. 1.1. The display family. The emissive family is listed by font. Displays with gray shade capability and analog segment displays are not considered here. The passive displays have the same font choice as the emissive ones; they are listed by type. Electromechanical displays are not considered further.

Table 1.1.   Light-Emitting Processes Available for Emissive Displays

Effect	Realization
Cathodoluminescence (CL)	VF
	CRT
Photoluminescence (PL)	Colored gas discharge (UV-excited phosphor)
	Fluorescent lamp based displays
Electroluminescence (EL)	VLED
	EL-excited, polycrystalline phosphor
Plasma decay	Gas discharge panel
	Nixie* tube
	Ac gas discharge
Blackbody radiation	Tungsten filament projection

*Nixie is a registered trademark of the Burroughs Corp.

Table 1.2.   Electrically Controllable Absorption Processes for Display

Process	Display realization
Dye or pigment (electrochromeric)	Electromechanical
	Electrochromic
Electropolarization	Nematic liquid crystal
	Field-effect liquid crystal
Electrophoretic	

Table 1.3.   Some Commercial Displays

	Fonts	Mechanism
Gas discharge:		
Dot matrix	Alphanumeric	Dc plasma
	5 × 200 Multirow	Ac plasma*,†
Planar flat panel	Multiple 7 segment plus decimal and symbol	Dc plasma
Raised cathode	Single 7 segment and decimal ½ digit 3 or 4 digit stackable	Dc plasma
Cathodoluminescent:		
Vacuum fluorescent	Multidigit 7 segment and decimal 5 × 7, 20 character 14 segment, multicharacter	Hot filament in vacuum tube with phosphor-coated segments
CRT	Selected character shape Character generation	Electron gun in vacuum with phosphor screen
Electroluminescent:		
Red GaAsP	Single- and multidigit 7-segment Alpanumeric	Monolithic or individual diode arrays
Red "Super Brite," orange, and yellow	Single digit 7-segment	Individual diodes and light pipe
GaAsP and green GaP	Alphanumeric	Individual 35-diode arrays
Passive displays:		
Liquid crystal	7-segment watch 7-segment calculator	Nematic or field effect
Blackbody or incandescent	Individual 7 segment and decimal	Hot filament(s) in vacuum

*Ac plasma not generally available.
†Photoluminescent when phosphor is added for color other than neon orange.

1.3   APPLICATIONS

In general, electronic displays will be used first to replace electromechanical displays in areas such as games, registers, automobile instrument panels, gas pump dials, TV dials, adding machines, cash registers, and counters. Second, they will be used to replace analog displays in applications such as D'Arsonval meters, clock and watch dials, scale dials, and temperature scales. Finally, they will be used for communications between digital machines and their operators, such as alphanumeric displays on prompting computers and "dialed" number verification on telephones. Table 1.4 lists some consumer and business applications and representative displays; a similar listing for other economic sectors is given in Table 1.5.

1.4   THE ELECTRONIC DISPLAY SYMBOLS

A minimum number of dots or segments (bars) are required to represent a given symbol set. In addition to typical patterns (Fig. 1.2) there are 10 segments for symmetrical 1s and plus signs, and 4 × 7 and 4 × 6 dot matrices for lower-cost dot

Table 1.4. Business and Consumer Display Applications

Application	Typical display
Appliance applications (washer, dryer, range, ovens, air conditioners)	
Time/time cycle Temperature/cycle	4 digit/Gas discharge, VLED
Commerce (adding machines, cash registers, scales)	
Totalizer/status Weight/unit cost	Multidigit with status indicator/VF, gas discharge, Tungsten filament
Automotive (clock, radio, instruments)	

Engine:
  Water temperature/oil pressure/vacuum
Peripheral:
  Time/frequency of radio, CB channel
Navigation:
  Gasoline remaining/mpg
  Distance remaining, distance to/travel
    distance
  Diagnostic readouts
	Gas discharge, VF, VLED

Personal consumer	
Watch Calculator TV, CB, radio	4 digit plus colon; VLED, LCD 8 to 12 digit; VLED, LCD, VF 2 to 4 digit; VLED, LCD, CRT

matrix displays. Sometimes a second small horizontal crossbar is supplied, which increases the 7-segment character to 8. Some typical character sets are shown in Fig. 1.2. Figure 1.3 shows some additional symbol sets available with the 7-segment display that may be useful as status indicators. Figure 1.4 shows a symbol set generated with 7-bit logic and a 5 × 7 display.

Table 1.5. Industrial, Medical, and Military Display Applications

	Typical display

Medical:
  Digital thermometer
  Sphygmomanometer, pulse rate, patient monitors
Industrial electronics:
  Meters, positioners
  Test equipment, gages
Military:
  Situation indicators
Miscellaneous:
  Computer peripherals
  ALU status
	
	VLED
	—
	
	VLED, tungsten filament
	Gas discharge
	
	Traditional
	
	VLED
	VLED

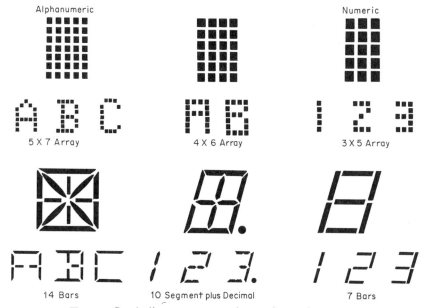

Fig. 1.2. Symbolic array geometries. *(After Ref. 1, in part.)*

Manufacturers may select characteristic fonts. Examples include square corner, round corner, and mitered corner (Fig. 1.5). Also shown in the figure is the standard segment identification method: segments are lettered from *a* through *g* and sometimes *h*. Characters may be upright or slanted; a ten-degree slant is common. Use of the slant has the advantage of allowing the decimal point to fit in the character space.

An additional font requirement is the aspect ratio, which is the ratio of height to width of either the character or the segment. For example, segment aspect ratios might be 5:1 for a vacuum fluorescent display and would result from phosphor deposition requirements. For a gas discharge display the segment aspect might be 10:1 or 20:1, where the display would be expected to appear crisper or sharper because of the higher aspect ratio. Character aspect ratios are in the neighborhood of 2:1; values smaller than 2:1 are often used in conjunction with a small aspect ratio or "fat" segments, and values larger than this are useful where multidigit space is at a premium.

Display size depends critically on the display technology (Fig. 1.6); various technologies are restricted to particular size ranges. Commonly used sizes are

Fig. 1.3. Some symbols available from the 7-segment character.

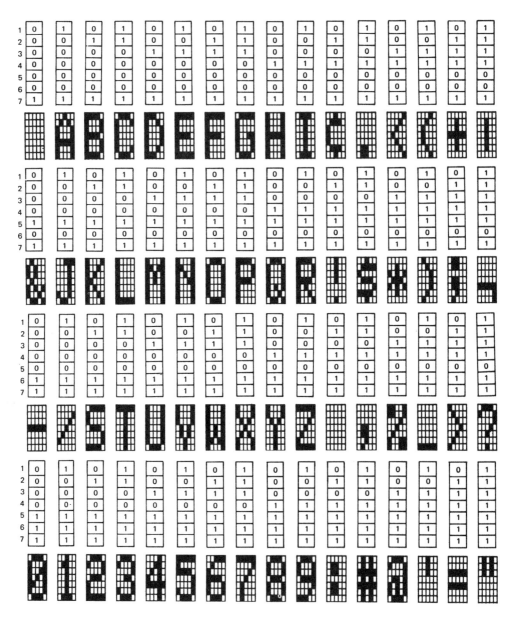

positive logic: 1 = H = 2 V to 5.5 V
0 = L = 0 V to 0.8 V

Fig. 1.4. 5 × 7 alphanumeric display type TIL305. Resultant displays using TMS4179JC or TMS4179NC chips with EBDIC coded inputs. *(From Ref. 2.)*

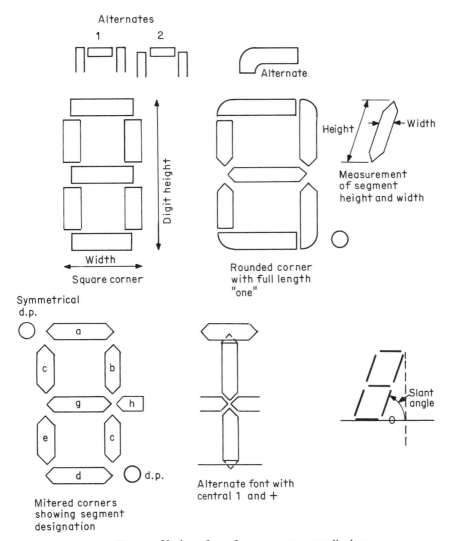

Fig. 1.5. Various fonts for seven-segment displays.

[centimeters (inches)]: 0.076(0.03), 0.15(0.06), 0.38(0.15), 0.46(0.18), 0.51(0.2), 0.76(0.3), 1.0(0.4), 1.22(0.5), 2.0(0.8), 2.54(1.0), 5.1(2.0), and 10.2(4.0).

A display may be limited to a single character, or it may use stackable combinations of single characters to form a larger number of characters. Stacking is common practice for VLEDs and raised cathode gas discharge displays. Other technologies with large fixed-cost packages find it more economical to fabricate large-character-number displays as a single package. Also, quantity applications invite the same procedure for the VLED technology. Some multidigit forms are $3\frac{1}{2}$ and $4\frac{1}{2}$ digit (where the $\frac{1}{2}$ digit is a ± 1). Other styles are 8 digit(d), 9d, 10d, 11d, 12d, 13d, and 14d, where the last odd-numbered character is generally used as a status flag. Other special symbols are shown in Fig. 1.7. Figure 1.8 shows a clock display. Morning or

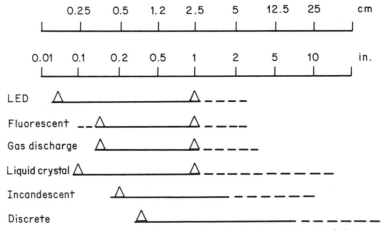

Fig. 1.6. Comparison of character sizes by display. *(After Ref. 3.)*

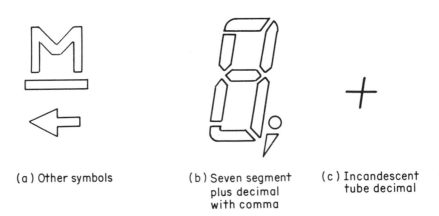

(a) Other symbols

(b) Seven segment
plus decimal
with comma

(c) Incandescent
tube decimal

(d) Other special legends and
symbols

Fig. 1.7. Special display symbols.

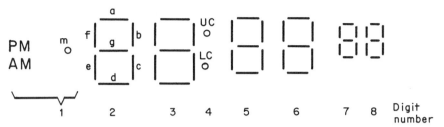

Fig. 1.8. Clock display. Abbreviations: UC—upper colon; LC—lower colon; m—moon; a, b, c, . . .—segment identifiers. The relative segment areas are moon:large segment:small segment:colon = 3:10:4:3. Either 6 digits plus colon and symbolization may be used for the clock or only four digits.

evening is shown with AM or PM symbols, or sometimes with a sun and moon symbol. To keep segment drivers the same, the segment and symbol area can be kept the same throughout a given display. Brightness per unit current density is translated into equal segment brightness at the same drive current per segment under such restrictions. This rule is also kept for the decimal and minus signs. In Fig. 1.8 this is not the case; the moon, colon, large segment, and small segment areas are in the relationship 3:3:10:4, and the drive currents must be scaled in the same ratios.

REFERENCES

1. Adapted from A. A. Bergh and P. J. Dean, *Proc. IEEE,* **60** (1972), p. 156.
2. Texas Instruments, Inc., Dallas, Texas.
3. E. G. Bylander, "Components for Electro-optics," in C. A. Harper (ed.), *Handbook of Components for Electronics,* McGraw-Hill, New York (1977), p. 5-68.
4. A. H. Agajanian, "A Bibliography on Display Technologies," *Proc. of the Soc. for Inf. Disp.,* **14** (1973), p. 76.

2

Electronic Display Fundamentals

2.1 SPECIFYING DISPLAYS

Introduction

The equipment designer controls the display manufacturer's product by means of the display specification. The display parameters that are amenable to specification and control are of first concern. Also of concern are unspecifiable properties that are subject to other kinds of controls. These parameters form the basis for comparisons between various kinds of displays or display technologies that help the designer select one from the many. Some qualities controlled by specifications are pleasing appearance, reliable operation, cost effectiveness, efficiency, ease of application, and freedom from electrical surprises (i.e., no defects that are not mentioned by the manufacturer should show up under operation extremes).

The equipment manufacturer controls the unit-to-unit display variations by means of the display specification. Some goals are high quality, little rework, high reliability, customer appeal, ease of testing, and costs on a learning curve.

The user controls the utility of the product, in part, by means of the display specification. Some goals are pleasing appearance, reliability, low fatigue factor, and reasonable equipment cost (i.e., no display cost penalty).

It can be seen that while some of these requirements are overlapping, others are specialized to one group or another. For example, the user is not interested in the ease of manufacture, except possibly through the no-cost-penalty requirement.

Further, while some factors cannot be quantized, many must be. For example, when the systems engineer states the display specification as one that the display must have a pleasing appearance ("You know, it must look good"), the display designer must translate this requirement into data sheet requirements. How is this to be done? First, the display is observed by people; people have differing viewpoints. It is said that there is no accounting for taste, and to the extent that members of the equipment team may say, "We don't like it, redo it"; this is true. However, to the extent that specifications must be written around a component that interfaces between the equipment and users in order to best fit their needs, taste can be taken into account. Human engineering makes this feat possible. The fields of human engineering and of displays are both new enough that many component design factors are only poorly understood or incompletely specified. Table 2.1 tabulates some display parameter areas and subdivisions. A judgment as to their specifiability is included. Chapter 3 will be concerned with bridging the gap between subjective requirements and measurable specifications. Areas that are not yet amenable to such control may be judged in two ways: (1) Ask salespersons at nearby retail outlets if their customers think the parameter is important.[1] (2) Give a working model, which incorporates the display and the parameter display to be tested, to unsophisticated users such as children or workers from an unrelated area and listen for their comments.

An example of the first method occurred in judging the importance of display color. It was determined by talking to salespersons that only one person had ever asked for a specific color display in the product; this person wanted red rather than green. The second method was also used. Samples with both orange and green displays were lent to office users on request. Female users who had orange display products on their desk invariably said that they really preferred green; in some cases it proved difficult to get the green display product returned. Most male users did not have a preference; those that did were about half-and-half orange or green.

An example of the second method involved a controversy concerning two different types of displays; an advantage of gas discharge displays over VLED displays was said to be visibility outdoors in sunlight. Two boys playing in the back seat of a car with a product utilizing the gas discharge display were overheard to say, "When you hold this under the back window, you can see only 8's." This remark is interpreted to mean that sunlight reflected off the unlit segments appears as the lit plus sunlight-reflected segments, and clearly rules out any advantage of the gas discharge display in outdoor viewing applications. Method 2 may be summarized as the "Emperor has no clothes method," in that many display myths have long lives simply because, although plausible, they are not tested.

Early, it will be important for the designer to decide minimum display quality

Table 2.1. Some Display Parameters Judged to Be Specifiable and Nonspecifiable

Conventionally controlled by specification		Specifiable		
Yes	No	Yes	No	Parameters
	X	X		Crisp, sharp, fuzzy, eye fatigue
	X	X		Aspect ratio
	X		X	Cool; hot (color preference by user sex)
	X	X		Dominant wavelength
	X			Color match
X		X		Brightness match
X		X		Imperfections
	X			Distortion as a function of angle
	X			Error rate by font (% of characters mis-read as other characters)
	X		X	Pleasing appearance
	X			Color fatigue
	X	X		Font character shape

specifications and exercise strong control. One reason is that most system designers and other technical customers are too severe on blemish specifications and must be sold on cost-effective blemishes that will not in any case interfere with the use or enjoyment of the display. A second reason is that these same customers are often too lax on display brightness. Use this rule of thumb: "If the display doesn't look bright enough, it isn't."

Major Design Factors

Gordon and Anderson[2] list three display parameters which are said to be important from a user's point of view: resolution, screen size, and bulk. Translating the first two requirements from image displays to character terminology would yield the font aspect ratio and digit size (based on the number of simultaneous viewers). These requirements must be expanded to include (assume a displayed brightness of 50 fL or a 10:1 contrast ratio[3]) cost per digit, including drivers, connectors and mounting hardware, power per digit, viewing angle, color, and environmental range outside of +10 to +55°C. These factors are, of course, added to those mentioned earlier. These design factors may be lumped under some major design categories:

- Design-to-cost
- Design-to-size
- Design-to-power
- Design-to-reliability

Two ways of allocating the display's portion of system goals will be mentioned. One is to use industry practice. This method is especially useful when the display cost is a small fraction of the system cost. For example, one dollar per automobile is a rule of thumb for many such components. The other way is to allocate the display's portion of its fraction of the overall system. As an example, suppose the display contributes one-fourth of the subsystems to a calculator (the others are the power supply, the chip, and the keyboard). Then, assuming a design-to-reliability goal of less than 3% return rate, the display's portion should be allowed up to ¾% as a contribution to the return rate, if necessary, in determining reliability criterion.

2.2 COMPARISON METHODS

Quite often it is necessary to compare various display technologies as a prelude to a choice between them. These comparisons will occur at different points of the design cycle (Table 2.2). It can be seen from the table that comparisons have different motivations, depending on the point at which they are made. For instance, during the predesign phase little is known about the specifics of any one technology, and there may be little to recommend one over the other. Here the comparison is done on the parameters discussed in the previous section. In the production phase, on the other hand, a great deal is known about the device in production and the proposed improvement represents a major unknown. There the comparison will be between well-known advantages and disadvantages of the production product and a relatively unknown new or improved version.

Table 2.3 is one example of a comparison of digital display technologies at a preproduction phase. Here both active and passive displays are compared. Note

Table 2.2.   Design Points of Comparison Between Display Technologies

Time of comparison	Use of comparison	Available history
Predesign	Selection of one from major category of displays.	Little. Other products may not be available for comparison.
Design review	Compare two major display technologies as contenders or compare improved version with earlier design.	Circuit tests. Initial procurement tests.
Preproduction	Review earlier decisions.	Life tests. Circuit variations pilot runs.
Production improvements (proposed)	Improved display or display circuit versus old.	Return rates; acceptable quality levels (AQLs); field reports; standard tests
Product improvement	New candidate versus old.	Relative costs, reliability, quality, power, and other parameters.

that, in addition to display power, power supply voltage is an important parameter. Again, in Table 2.4 the Self-Scan* is compared with the cathode-ray tube. Important parameters are seen to be data capacity versus operating voltage and bulk. A comparison between the liquid crystal (LCD), electrochromic (ECD), and the electrophoretic (EPID) displays is given in Table 2.5. Note that many of the parameters included in Table 2.3 are omitted here. Since many manufacturers' data sheets do not discuss ruggedness or ease of mounting, for example, this data may need to be obtained from engineering samples.

A comparison made during the predesign phase of a portable product might consider cost, reliability, power consumption (battery life), and appearance. A power-consumption comparison between two production models of a popular calculator is shown in Table 2.6. In the actual comparison, there proved to be insufficient reliability and cosmetic data available on the two display types to permit a comparison, but the power-consumption comparison was overwhelmingly unfavorable to the gas discharge display. In another comparison (Table 2.7) between a portable calculator with production VLED or proposed production improvement gas discharge display which had much larger characters, there appeared to be no difference in power consumption. The offsetting factor was increased component count and cost for the gas discharge display. Finally, a design-stage comparison of display costs for a 10-digit desktop calculator is given in Table 2.8. (All costs are referenced to the MOS cost as 1.) It is seen that there were only small differences in cost between the technologies at the time.

2.3   ADDRESSING METHODS

The addressing may be in a parallel mode or through some combination of serial and parallel modes. The latter method is known in the electrooptical systems business

---

*Self-Scan is a registered trademark of the Burroughs Corp.

Table 2.3. Preproduction Comparison of State-of-the-Art Display Technologies

	No. of digits/ module	Cost[a] of display only ($/digit)	Cost[a] of decoder/ driver ($/digit)	Total cost[a] of integrated display ($/digit)	Character height (in)	Character style[e]	Color(s)	Maximum viewing distance (ft) Indoor	Maximum viewing distance (ft) Outdoor	Maximum viewing angle (degrees)	Display voltage[b] (volts)	Figure 8 power[c] (watts)	No. of digits which can be multiplexed	Life (hours × 10⁵)
Numerics:														
Gas discharge	1–32	1.48–3.75	0.42⁵–1.50	2.02–5.83	0.2–0.55	7-S, FF	Neon orange	15.40	8–20	60–65	160–250	0.005–0.562	10–16	1–2
Incandescent:														
Directly viewed	1–20	1.95–11.00	1.00–5.75	2.95–16.75	0.38–0.71	7-S	Any	20–35	5–32	50–70	2.5–5	0.2–0.755	8–16	0.5–2.5
Backlighted	1–9	4.50ᵈ–19.00	4.50–11.00	10.00–26.00	0.32–3.00	7-S	Any	10–85	10–40	60–85	2.5–2.8	0.72–0.9	None–all	0.05–1
Rear projection	1–1	14.40–76.00	No data	No data	0.38–3.38	FF	Any	No data	No data	No data	5–28	No data	No data	No data
Vacuum fluorescent	1–13	1.35–2.77	1.50–2.00	4.25–4.55	0.18–0.57	7-S	Green plus some filters	8–40	5–8	60–75	20–50	0.07–0.172	8–16	0.1–1.5
Cathode-ray tube	1ᵈ	22.50	No data	No data	0.62ᵈ	FF	Green, red, blue	No data	No data	No data	Specialᵈ	No data	No data	No data
Light-emitting diode	1–6	2.95–18.75	0.35⁵–10.00	3.78–27.50	0.1–0.77	7-S, DM	Red, green, yellow	3.3–49	4–40	45–90	1.6–5	0.048–1.25	2–100	1–infinite
Liquid crystal	3–8	2.00–5.00	1.43–9.12	6.27–12.10⁴	0.18–0.65	7-S	White or filtered	8–30	8–30	70	12–60	>0.001–0.001	0–4	0.1
Electroluminescent	9–21	1.05–2.10	No data	No data	0.25–0.45	7-S	Yellow-orange	15–25	5–10	80	650	0.096–0.15	50	0.25
Alphanumerics:														
Gas discharge	16–80	3.91¹–11.00¹	No data	3.91¹	0.2–0.4	DM	Neon orange	No data	No data	No data	200–250	0.278–0.8	16–80	No data
Incandescents:														
Directly viewed	1–1	8.10–22.00	45.00–	56.40–67.00	0.31–0.62	16-S	Any	15–25	3–21	50–60	5	0.28–0.63	12	1–2.5
Backlighted	1–any	23.00–32.00	No data	32.00	0.32–0.75	16-S	Any	10–15	10–15	75	4.5–5	0.72–4.8	None–all	0.05–1
Rear projection	1ᵈ	91.57	No data	No data	0.45ᵈ	FF	Any	No data	No data	No data	5ᵈ	No data	No data	No data
Cathode-ray tube	1ᵈ	55.00	No data	No data	0.62ᵈ	FF	Green	No data	No data	No data	Specialᵈ	No data	No data	No data
Light-emitting diode	1–5	10.00–39.00	11.50⁵	21.50–39.00	0.27–0.35	DM	Red	9.2–26	8	60–75	1.2–12	0.084–2.55	10–any #	1–10
Electroluminescent	9ᵈ	4.00	No data⁶	No data	0.45ᵈ	13-S	Yellow-orange	20	7	80	650	0.144	50	0.25

ᵃ1,000 quantity, unless otherwise specified.
ᵇSome displays require several voltages. Voltage listed is for light-emitting portion.
ᶜTotal power for all supplies.
ᵈOnly one display reported.
ᵉ7-S, 7 segments; FF, fully formed; DM, dot matrix; 16-S, 16 segments; 13-S, 13 segments.
¹1,000 quantity
²100–999 quantity
³1–24 quantity
⁴Combined cost of 2 and 3.
⁵Some displays include cost of decoder-drivers.
⁶No standard IC. Must be made from discretes.
SOURCE: Ref. 4.

Table 2.4. Preproduction Comparison of the Cathode-Ray Tube and Self-Scan* Dot Matrix Tube

	CRT	Self-Scan II
Life expectancy	20,000 hours	50,000 hours
Character readability	Acceptable under certain operating conditions. More difficult to read in most applications.	Excellent under all operating conditions. No fuzziness, splash, smear. Readable from long distances.
Legibility	Characters uniform ± 10 or 20% except at display boundaries.	Distortion free, extremely uniform characters.
Driving voltage	High voltage, X-ray hazard.	Medium and low voltages.
Message display area	10 in² (5-in tube)	20.6 in² (480-character panel)
Character height	Variable. Most commonly, 20 in.	0.26 in. for 240- and 480-character panels.
Cost	Slightly lower.	Small cost advantage in favor of CRTs.
Resolution	Very high.	To 60 lines/inch.
Light output	Very high light output (with limited life).	Bright without eyestrain.
Effect of outside influences	Affected by magnetic fields.	Not affected by magnetic fields.
Size	Bulky, inconvenient size especially where space saving is important.	Compact, flat panel.
Data capacity	High data density; often much unused capacity.	To 480 characters or more; acceptable for many applications.
Weight	Up to 50 times heavier with electronics	Low
Maintenance	Periodic maintenance required to adjust focus, pin cushioning and centerings.	No periodic maintenance required; no pin cushioning or focus problems.

*Registered trademark of the Burroughs Corp.
SOURCE: Reference 5.

as *multiplexing* (MUXing) and in the optoelectronics components business as *strobing*. It has the advantage of reducing the number of interconnections from d times s (d = digit, s = segment) to d plus s connecting leads in typical cases. This option is particularly attractive to chip designers, who are severely restricted by lead number. Smaller numbers of interconnections also lead to reduced assembly costs and lower chance for failures. This option has the disadvantage of reducing the apparent brightness by $1/N$, where N is the number of strobed elements. A display with limited brightness will be eliminated from competition if a strobing requirement is added. Displays must be strobed at a rate which exceeds the fusion frequency of the eye (Chap. 3). Slow-response-time displays may be difficult to strobe for this reason.

Seven-segment displays commonly have all like segments tied together and operate with multiplexed digits. A 10-digit display with an 11th character would be strobed at a duty cycle of $^1/_{11}$ or $^1/_{12}$ where the latter would apply if a 10% blanking were used between digits [duty cycle equals digit on-time divided by total number of equivalent digits (cycle) time]. A 5×7 dot matrix might have strobed columns with

Table 2.5. Summary of LCD, ECD, and EPID Characteristics

Category	Comments		
	LCDs	ECDs	EPIDs
A. Visual appearance	1. Medium viewing angle. 2. Contrast is viewing-angle sensitive. 3. Many colors are available.	1. Very wide viewing angle. 2. Contrast is not sensitive to viewing angle. 3. Contrast is limited because of operating-life constraints. 4. Some color selection.	1. Very wide viewing angle. 2. Contrast is not sensitive to viewing angle. 3. Contrast can exceed 30:1. 4. Wide range of colors.
B. Power dissipation	$1 \mu W/cm^2$–$1.0 \ mW/cm^2$ at 3–15 V ac.	1. 4–10 mJ/cm² per switching cycle at 1.0 V dc. 2. Power dissipation = 4–10 mJ/cm² × number of switching cycles per second.	1. 20–100 $\mu J/cm^2$ per Switching cycle at 30–100 V. 2. Power dissipation = 20–100 $\mu J/cm^2$ × number of switching cycles per second.
C. Reliability	>10–20,000 hours	$>10^6$–10^7 switching cycles	$>10^6$–10^7 switching cycles in 3000 hours
D. Response times at 20°C	10–500 ms at 3–10 V.	1. 100–500 ms at about 1.0 V. 2. Hours to days storage.	1. 50–100 ms at 50–100 V. 2. Hours storage.
E. Circuit compatibility	1. Very low-current, low-to-medium voltage device. 2. CMOS IC compatible. 3. Bipolar waveforms necessary. 4. Multiplexing is limited (~4–8 lines).	1. High-current, low-voltage device. 2. Bipolar transistor circuitry more suitable than MOS. 3. DC levels with both polarities required. 4. Multiplexing capability is not known.	1. Low-current, medium-to-high voltage device. 2. Probably discrete transistors required for driving. 3. Low-frequency, bipolar waveforms needed. 4. Multiplexing capability is not known.
F. Temperature dependence	1. Temperature range of operating parameters is noticeable. 2. Operating range 0 to +70°C.	1. Temperature range of operating parameters is noticeable. 2. Operating range −20 to +70°C.	1. Temperature range of operating parameters is noticeable. 2. Operating range −15 to +50°C.

SOURCE: Ref. 6.

interconnected rows of dots and with adjacent characters driven in parallel. (Rows and columns may be interchanged in such schemes if timing requirements can be met.) Alternatively, all rows might be interconnected and columns in a $5 \times C$ column (C = character) display could be strobed. The number of columns is not only limited by the individual dot brightness, but also by the dot rise and fall times relative to the cycle time. When the duty cycle reduces the pulse width to less than their sum, the brightness will be reduced. Column driver requirements can be reduced for displays made up of bistable dot elements by use of multiphase clocking techniques (Chap. 4).

Table 2.6.   Production Comparison of VLED vs Gas Discharge Power Consumption

Calculator	Power	
	Worst*	Best†
Four-function hand-held VLED calculator	144 mW	96 mW
Four-function hand-held gas discharge calculator	312 mW⁺	312 mW
Change in power due to display difference	168 mW	216 mW

*All 8's.

†4 × 5's.

Conclusions: Two additional batteries required to offset increased power of gas discharge.

Alternatively, a display may be driven from a serial source in a dc fashion by means of a serial 4-bit or binary-coded-decimal (BCD) input to 7-segment output, for example. Such techniques allow the use of low-brightness or slow-display techniques. Specific examples for these techniques will be described in later chapters.

Restrictions imposed on matrix displays by element-to-element crosstalk are considered in Appendix 2.1.

2.4   CONNECTORS AND MOUNTING

Good mechanical practice is the heart of a good display design. While this subject will not be treated in detail here, some common practice will be discussed. Figure 2.1 shows a display mounted adjacent to a bezel which acts as a glare shield. Included as a bezel cover is a filter or contrast-enhancement media. The display itself is attached to a rigid mount, which may be a printed-circuit (PC) board or other structural member, by means of an elastic adhesive material that accommodates differences in thermal expansion between the display and its mount. Alternatively, the display may be attached by means of its electrical leads, provided that the display is spaced to allow for expansion and contraction differences between it and the mount, and the leads are specified to have sufficient strength for the application. Where long flexible leads are provided, as opposed to pins, provision for stress relief is good practice.

Often individual display assemblies are made up on individual PC boards. In this

Table 2.7.   Production Improvement Comparison: Battery Life of a Scientific Calculator

14-Digit display	Total power		Best hours
	Best	Worst	
VLED	375 mW	619 mW	6
Standard gas discharge	384 mW	575 mW	6

Conclusion: Standard gas discharge is competitive with VLED.

Table 2.8.   Design-Stage Comparison of 10-Digit Desktop Calculator Display Costs*

Option	Display associated parts						
	Display plus connector	ICs	Diodes	Cap	Transistors	Resistors†	Cost total
Gas discharge (direct drive)	1.10	0	0.14	0.15	0.26	0.06	1.71
VLED	1.83	0.27	0.04	0.06	0.04	0.04	2.22
Fluorescent	1.15*	0	0.06	0.08	0.04	0.01	1.34

*Costs relative to MOS chip = 1 and rounded to two places.
†Excluding pull-down resistors.

case the PC board can be connected to the mother board by means of pluggable contacts or a lead frame interconnect to a mother board (Fig. 2.2).

For reasons of reliability, integral or hard contact systems are to be preferred for interconnecting displays to circuit boards. However, where pluggable connectors are to be used, design goals should be (1) wiping action, (2) soft metal to metal, and (3) use of stress relief or floating connector element to maintain stress relief during thermal cycle or on-off cycle to less than contact adhesion. Additional connector considerations may be found in Ref. 7.

2.5   GENERAL TEST METHODS

Test Philosophy

It will be desirable to test the parameters discussed to determine the compliance of an individual unit to set specifications, or to intercompare display types. The general test requirements to accomplish this goal will be discussed here. Testing

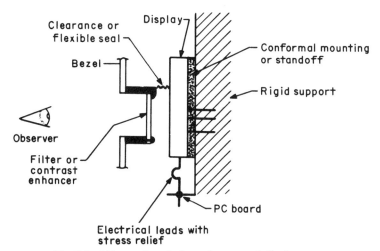

Fig. 2.1. Side sectional view of a general display mount.

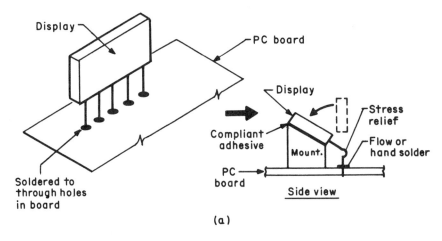

(a)

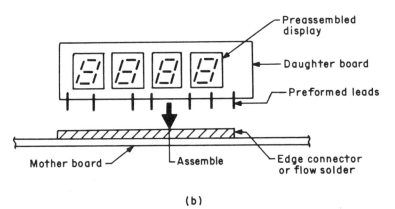

(b)

Fig. 2.2. Two alternate display mounting procedures. (*a*) Display leads soldered to PC board; (*b*) mother board approach.

may be done in the actual product, in a special test set, or in the burn-in rack. All these techniques have offsetting advantages and disadvantages. Testing in the product may make it difficult to obtain individual parameters, or it may just provide a general operational test. Also, it may lead to testing drive-circuit flaws rather than display performance. On the other hand, a special test circuit may allow all parameters to be determined but fail to reveal operational weaknesses that may show up in the actual performance. The best approach may be to do the manufacturing testing on the burn-in rack and quality lot acceptance in the customer's circuit of final use. Engineering tests may be done on all three types of circuit.

Strictly speaking, test sets for life tests or environmental tests must be the same as the circuit of final use. However, to develop failure modes or to develop failure models, alternate lower-cost approaches may be necessary. Accelerated life testing will involve increasing one or more parameters to cause the onset of previously determined failure modes, and such testing may also require some values of drive to be outside the standard circuit range. These values can be furnished on burn-in racks as modified for accelerated life testing or similar circuits.

Test Parameters

Testing breaks down naturally into the following categories: cosmetic tests, environmental tests, reliability tests, brightness tests, and electrical tests.

Reliability Testing.   To test the reliability of a device, it is necessary to determine its life in a specific application (on the average). The life of a device will be determined by an end-of-life criterion and the time it takes to reach this point. An end-of-life definition can be that the device no longer meets the incoming specification or it can be based on a predominate failure mode for the device. A device that fails by loss of light output is conventionally defined to fail when it reaches half brightness. Another end-of-life criterion is "still readable." An end-of-life definition based on failure mode implies an understanding of dominant failure modes for the device of interest. There are at least three ways to develop failure modes: in-service tests, in-circuit life tests, and accelerated testing involving an increased stress which is usually temperature and/or current. Various tests may be based on convention or on guidance afforded by various military standards.[8]

There will be several possible goals of life testing: develop failure mechanisms, correlate failure mechanisms to accelerated test methods, and develop failure models. Failure mechanism studies lead to guidance for the designer to improve the device, to the production engineer to negotiate reliability specifications, and to the test engineer to devise meaningful tests. Accelerated test methods are important to quickly develop improvements and to monitor the continuing product quality. Failure models are the basis for many design to reliability efforts.

At least three questions are of importance in life test planning. The first is the number of test units: one is usually too few, and a larger number is a burden in drive circuit and wiring cost. Numbers between twenty and two hundred usually prove to be practical. Twenty allows one to take into account an occasional nonrepresentative failure without losing the test and is still small enough to allow wiring in under three weeks' time. Two hundred units allow accumulation of a large number of total test hours in a short time, but at $50 to $100 per display-circuit cost, the total cost may be prohibitive. Large numbers of test units have proved most convenient for rolling life tests where samples of weekly production lots are rolled on (and off) life test in 20-unit lots; and for final qualification tests of new displays or of displays with major innovations.

Environmental Testing.   Environmental tests are usually based on a selection of tests described in various military standards.[9] Two kinds are of interest: the effect of environmental stress on brand new parts, and the effect over life. The procedure for selecting tests is similar to reliability testing: first, a definition of failure is accomplished and may be based on failure modes or on the incoming specification. Worst-case designs, worst-case power supply, and worst viewing conditions must be defined or determined. Circuits are selected in the order of preference mentioned earlier: in service, circuit of use, or test circuit. Tests are designed based on goals similar to those above. Such goals may be to develop correlation with field failures to determine the particular source, to develop a failure model, or to improve the design so as to increase stress resistance.

Some recurring problems in display testing are the following. Determining the temperature rise of an operating display installed in a product is one that continually recurs. This rise may easily be determined by attachment of a thermocouple; the usual result is a $+3$ to $+5°C$ change. Another recurring question is the amplification of vibration amplitude between the product and its display mounted inside it, and also the applicability of stand-alone vibration or shock tests done on the display to in-service performance. In practice, this question is resolved by in-service, or "shake, rattle, and roll," tests which show no failures attributable to the display.

A recurring question involving temperature tests is: Should the performance be determined over the entire range? In practice, such tests are not worth the cost; tests at ambient and at the extremes are usually sufficient. Many times the display will not be operable at one or both extremes; in this case and some others above, it *must* be remembered that *no* amount of testing will improve a display or its reliability or stress resistance. The display must be redesigned or some circuit solution sought to solve the problem, or the problem may be insolvable in the sense that resources are not sufficient to overcome the difficulty. Then an adaptation to the problem must be made. Sometimes the old adage, "If you have a lemon, make lemonade" will allow a resolution. One case of a display failure mode at a current extreme within the specification limits required a circuit modification. This change also lowered the drive chip burn-in failure rate from 3 to $\frac{1}{2}\%$ and led to substantial manufacturing savings at no noticeable performance loss.

Electrical and Cosmetic Testing. Electrical tests are important to determine display performance as well as to allow monitoring of the above tests. The test circuit can be a modified production circuit with a pinned-out display connector. Quick-disconnect sockets are particularly useful for this purpose (Fig. 2.3). Tests can also be performed on the burn-in rack, where such a procedure can eliminate one display loading and unloading operation. A disadvantage is that such tests may miss shorts and opens.

A convenient test circuit block diagram is shown in Fig. 2.4. This circuit can allow diagnostics by measuring segment or element current individually, for example, by measuring shorts and opens. Parameters of units cycled off life test can be determined to be in specification range.

Typical display tests are to drive the display at extremes based on data sheet

Fig. 2.3. An example of a quick disconnect socket for display testing. *(Courtesy of Textool; photo by Quin.)*

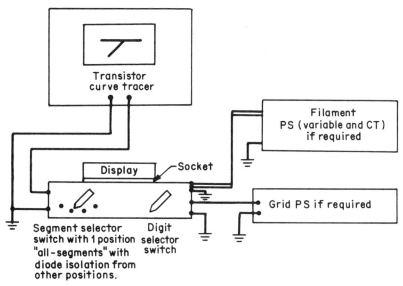

Fig. 2.4. A general display tester (e.g., V.F., gas discharge, VLED, LCD). It is useful for characterization and diagnostics.

typical operating parameters or absolute maximum ratings or extremes determined from in-service experience. A low-current test and a high-current test are typically used. The unit is tested at both extremes for cosmetic blemishes against the allowed limits with all eights set into a bar display or all dots of a dot-matrix display. Then 1, 2, 3, 4, 5, . . . , are entered to detect shorts and opens. These tests are usually sufficient for most purposes.

Photometric and brightness tests are discussed in the next chapter, but cosmetic testing by trained operators is surprisingly precise. Typical tubes can be retained for standards, comparison, training, and refreshing purposes, but in general, subjective judgments are reproducible. Ten to fifteen seconds is a typical test time for a 10-digit display.

REFERENCES

1. Technique communicated by William P. Stearns.
2. E. I. Gordon and L. K. Anderson, *Proc. IEEE,* **61** (1973), p. 807.
3. As a rule of thumb, displays that do not meet this requirement will not be on the market.
4. Anonymous, *The Electronic Engineer,* August 1972, p. 45.
5. Burroughs Corp., East Plainfield, N.J.
6. L. A. Goodman, *Proc. of the Soc. for Inf. Disp.,* **17** (1976), p. 30.
7. B. R. Schwartz, "Connectors and Connective Devices," in C. A. Harper (ed.), *Handbook of Components for Electronics,* McGraw-Hill, New York (1977), pp. 11-1–11-139.
8. See, for example, Saul Zatz in Ref. 7, pp. 1-12–1-39; B. R. Schwartz in Ref. 7, pp. 11-105–11-124.
9. Display reliability in a standard environment is usually determined first; only later are reliability specifications established for harsh environments. See, for example, Ref. 8.
10. Textool, Inc.

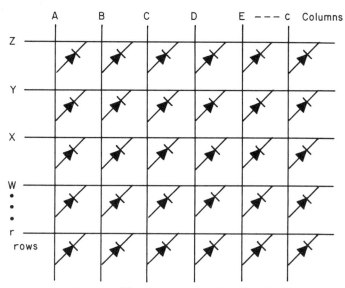

Fig. A2.1. Display matrix. *(After Ref. 1.)*

APPENDIX 2.1
Multiplexing of Matrix Displays*

When an array of display elements is arranged in rows and columns as in Fig. A2.1, there will be a limit to the number of columns of a given row that can be excited without introducing crosstalk. If columns A and B are driven and row Z is selected, Fig. A2.1 may be redrawn as Fig. A2.2.

Suppose all the elements are symmetrical and have an i-v curve as shown in Fig. A2.3. It is assumed, of course, that the elements are either on or off for voltages below V_{th} and otherwise above V_{th}. Then the voltage V_{AZ} between A and Z cannot be increased beyond some value if V_{EZ} is not to exceed V_{th} of device EZ:

$$V_{AZ} < \frac{s(r-1)+c}{s(r-1)} V_{th} \qquad s \neq c$$

where r = number of rows
$\quad c$ = number of columns
$\quad s$ = number of selected columns

As Sobel observes, the applied voltage can only be 25% more than V_{th} if $s = 4$. The restriction above is not as severe for nonlinear devices such as light-emitting diodes.

To derive this result, Fig. A2.2 is further decomposed; Fig. A2.4a is for s odd; Fig. A2.4b is for s even. The voltage V_{Az}, V_{Bz} is selected equal to V_1, and by symmetry the nodes V_2 and V_4 are equal. For the case of odd s, $V_2 = V_3$ since i_{23} (current from node 2 to node 3) is 0:

*Based on A. Sobel, *Proc. IEEE, Trans. Electron Devices,* **18** (1971), p. 797, and a circuit description attributed to E. E. Loebner.

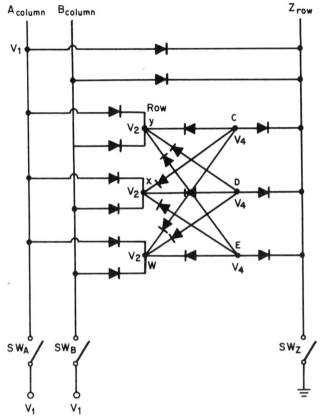

Fig. A2.2. Figure A2.1 as redrawn. *(After Ref. 1.)*

$$i_{23} = 0 = (V_2 - V_3)R_{23}^{-1}$$

Alternatively, note that the currents from node W to reduced common node 4 through R must equal the current out of node W through an equal equivalent R to node 4.

Now from Fig. A2.4c, $V_1 = iR_t$ and $V_4 = iR/(c - s)$. Then for the odd case:

$$\frac{V_1}{V_4} = \frac{R_t}{R/(c - s)}$$

where R_t has been shown to be equivalent to the even case ($R_2 = R_3$) and for this case.

$$\frac{V_1}{V_4} = \frac{\dfrac{R}{s(R - 1)} + \dfrac{R}{(c - s)(r - 1)} + \dfrac{R}{c - s}}{\dfrac{R}{c - s}}$$

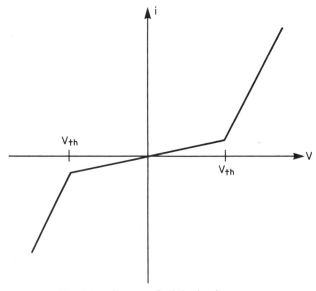

Fig. A2.3. Symmetrical device i-v curve.

which reduces to

$$V_1 = \frac{c + s(r - 1)}{s(r - 1)} V_4$$

Sobel shows that the use of clamping voltages can be used to relax this restriction. In fact, clamping removes the r, s, and c dependences. The result given when the maximum back bias is less than either V_{th} or an element reverse breakdown voltage V_{bd}, when this bias is used, is:

$$V_{appl} < 2V_{th} + |V_{bd}| \qquad \text{for asymmetrical elements}$$

or $\qquad V_{appl} < 3V_{th} \qquad \text{for symmetrical elements}$

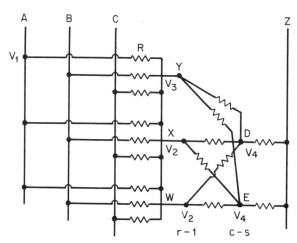

Fig. A2.4a. Unsymmetrical case for three selected columns.

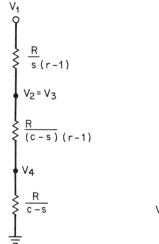

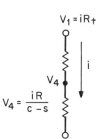

Fig. A2.4b. Reduces Fig. A2.2 for symmetrical display elements.

Fig. A.2.4c. Figure A2.4b reduced further.

If line-at-a-time multiplexing is used, only columns (or rows) need be clamped.

Clamping may be accomplished by using pull-down resistors; note that power dissipation is increased by this method. It is essential that both the clamp and the driver represent low-impedance sources during their operation; i.e., they must have good voltage regulation.

3

Display Human Factors

3.1 INTRODUCTION TO DISPLAY HUMAN FACTORS

For our purposes, human factors can be thought of as the study and optimization of the human-machine interface which seeks answers to the question "Can the machine be used?" or more broadly, "Is it of any value?" The typical answer is, "No, if people cannot use it long enough in their work location, or if they make frequent errors as a result of its use." Specifically, the question for the user-display interface is "Can the display be read easily, accurately, and without operator fatigue?" (Display factors that relate to this question are visibility, legibility, low fatigue factor, and lack of distracting elements. Specifiable quantities related to these factors are: to visibility are display brightness and contrast; to legibility are size, clarity, and font; to low fatigue factor are font, color, and absence of flicker; and to lack of distracting elements is freedom from defects.) Finally, limits may be placed on these quantities that optimize their human use.

When the above discussion is reviewed, it can be seen that a passage has been made from a requirement that the display be pleasing, to quantities such as size that in principle can be measured and correspondingly specified. These quantities will be discussed in this chapter.

3.2 VISIBILITY

Here will be considered two of the quantities which determine visibility: display contrast and brightness. The placement of two display bars on a white background (Fig. 3.1) illustrates the concept of contrst. The background is illuminated to give the apparent light level shown in Fig. 3.1b. When the bar or active bar has less contrast than the background, it is said to have negative contrast (level 1). When the bar has the same level as the background, there is no contrast (level 2); when it has a greater level (level 3), it has a positive contrast. Note that if the illumination source is turned down or moved away until its level equals level 1, the bar with level 2 brightness has a positive contrast (and also when it has level 3). Level 1 for the bar now yields no contrast, where before it was negative. This example shows the interplay of display and background brightness. Contrast may be enhanced by

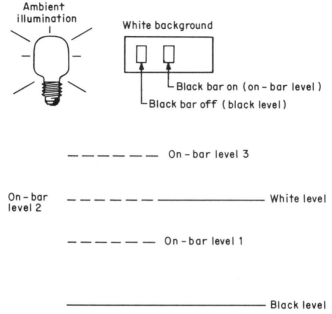

Fig. 3.1. (*a*) Black display elements with white background illuminated with an incandescent light bulb. (*b*) On-bar light level compared to background (white level) for three cases.

increasing the display brightness and by control or reduction of the reflected background light.

Definitions of contrast and brightness are considered next; then suitable values for contrast are recommended.

Definitions of Contrast

Some of the definitions for contrast and contrast ratio are given in Table 3.1. Note that source and background luminance must be in the same (or equivalent) units.

To illustrate this point and some of the definitions, a somewhat artificial example will be considered. In the example the display and the background will be chosen to have the same color.

Referring to Fig. 3.2, consider a two-bar display located in a background with a luminance of one lumen per square meter (1 lm/m²). If one bar is on and one is off, the contrast between the on-bar and the off-bar can be determined (bar-to-bar contrast), as well as the on-bar to background (display) contrast. For the first case, let the off-bars be shutters in the background: white matte when off and open to a source with an illuminance of 1 lm/m² [or 1 lux (lx)]. The source is to have the same color temperature as the background luminance of 2042 K.

The bar-to-bar contrast can be calculated as the normalized difference of light, reaching the eye from on-bar minus light from off-bar normalized to the off-bar.

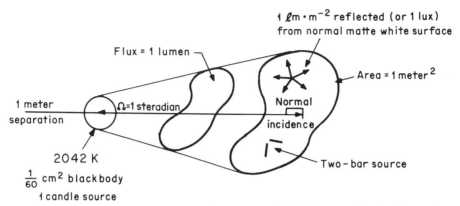

Fig. 3.2. Contrast example. A one candle (cd) source is incident on a Lambertian surface of area one square meter placed one meter away. The solid angle for radiation incident on the square meter normal to the background is one steradian. The incident flux is one lumen (lm), the flux density at the background is 1 lm/m², and the reflected light (apparent brightness of the background) is 1 lm/m², in any reflected direction (or $1/\pi$ cd/m² = 1 apostlib). A small two-bar source is located in the background.

The result is

$$\text{Contrast } C = \frac{1\text{ lx} - 1\text{ lm/m}^2}{1\text{ lm/m}^2} = 0$$

and the on-bar has no contrast with the off-bar. (Note that 1 lx = 1 lm/m².)

The on-bar–to–background contrast can be calculated in the same fashion as light from on-bar minus light from the background normalized to the background.

The display contrast is zero for the case under consideration. Now the off-bar shutter can be opened and its illumination source can be blocked so that no light is reflected from it.

The bar-to-bar contrast then becomes

$$\frac{1\text{ lx} - 0\text{ lm/m}^2}{0\text{ lm/m}^2} = \infty$$

or black-to-white contrast.

In a third case the on-bar can be a gas discharge in front of a matte white (Lambertian) cathode. Assume its characteristics simulate a 2045°C blackbody with 1 lx intensity. The light from the on-bar is now made up of three terms: the discharge, the discharge light reflected from the cathode, and the background illumination reflected from the cathode.

As noted earlier, 1 lm/m² is reflected from the one-candlepower source by a white matte finish; this result also holds for the cathode. The gas discharge source rearward radiation is equal to the forward radiation. In any practical case it will be close to cathode compared to the cathode width; therefore, an illuminance of 1 lx incident on a Lambertian surface will yield 1 lm/m² luminance.

Table 3.1. Definitions of Contrast and Contrast Ratio

Definition	Terms	Notes	Reference
1. $C = \dfrac{B_b - B_o}{B_b}$	B_b—Background luminance B_o—Object luminance	For objects darker than surround	a,b
2. $C = \dfrac{B_2}{B_1}$	B = any luminance $B_2 > B_1$	Contrast ratio known as contrast in TV engineering	a
3. $C_{min} = \dfrac{B_1 - B_2}{B_A}$	As above B_A = average	Minimum perceptible contrast. Reduces to (1) for equal area B_1 and B_2 so that $B_A = \frac{1}{2}(B_1 + B_2)$	a
4. $\dfrac{I_{max} - I_{min}}{I_{max} + I_{min}}$	I_{max} = diffraction pattern maximum I_{min} = diffraction pattern minimum	Laser	a
5. C	$\dfrac{L_1 + L_2}{L_2}$	L_1—Luminous flux from lit LED L_2—Luminous flux of background reflected from LED surface. Activated $\rightarrow$ nonactivated diode No cover	c
6. C	$\dfrac{(L_1 + L_2) - L_3}{L_3}$	(5) considering L_3, luminous flux of background light reflected from display background. No cover.	c
7. C	$\dfrac{(T_n L_1 + T_n^2 L_2)}{T_n^2 L^2}$	(5) with neutral-density filter as a cover (Transmission = T_n)	c,d
8. C	$C = \dfrac{BL_1 + SL_2}{SL_2}$	(5) with bandpass (BP) filter (cover), where B is lit source fractional output through filter S is background illumination of unlit source returned through BP filter.	c
9. C	$C = \dfrac{B'L_1 + S'L_2}{S'L_2}$	(5) with circular polarizer cover. B' and S' covered in reference.	c

Definition	Terms	Notes	Reference
10. $PC = \dfrac{\Delta I}{I} \times 100$	PC = percent contrast ΔI = (symbol luminance + I) − I I = background luminance	Apparently another form of (2)	e
11. Luminance contrast = $\dfrac{\text{segment} - \text{background}}{\text{segment}}$			f

References:

a. R. L. Kuehn and H. R. Luxenberg (eds.), *Display Systems Engineering,* McGraw-Hill, New York (1968), p. 106.

b. H. R. Blackwell, *J. Opt. Soc. Am.,* **36** (1946), p. 624.

c. J. Pucilowski, R. Schuman, and J. Velasquez, *Appl. Opt.,* **13** (1974), p. 2248.

d. W. D. Partlow, *Appl. Opt.,* **11** (1972), p. 1491.

e. C. E. Semple, Jr., R. J. Heapy, E. J. Conway, Jr., and K. T. Burnette, Tech report AFFDL-TR-70-174 (1971) Available as AD884770.

f. J. Duncan and S. Konz, *Proc. of the Soc. for Inf. Disp.,* **17** (1976), p. 180.

The on-bar–to–background contrast in this case is:

$$C = \frac{1\ \text{lx} + 1\ \text{lm/m}^2 + 1\ \text{lm/m}^2 - 1\ \text{lm/m}^2}{1\ \text{lm/m}^2} = 2$$

The bar appears twice again as bright as the background.

The principle applied to obtain the light at the eye from the second on-bar contribution (discharge light incident on white cathode) can be used to calculate the contribution of an extended diffuse source to the background luminance as opposed to a point source (such as a table lamp) considered earlier. If the room has a diffusely lit ceiling and the walls are Lambertian, the flux can be considered uniform. If a light meter measures this flux as 100 lx, then a white diffuse display background will have a uniform luminance of 100 lm/m^2.

Table 3.2 summarizes various sources of background illumination.

Photometry

Photometry, or the measurement of apparent brightness, is based on a standard detector and a standard source. The standard detector is the standard eye response or 1931 CIE (Commission Internationale de l'Eclairage) curve (Fig. 3.3). The standard source is the standard candle which is based on a blackbody of $1/60$-cm^2 aperture held at the temperature of freezing platinum (2045 K). This source has a luminous intensity of 1 candela (cd) (or also one lumen per steradian).

A curve such as Fig. 3.3 is a detector spectral response curve and is represented as $K_\lambda d\lambda$ or the response per unit wavelength interval.

A 2042-K blackbody spectra is shown in Fig. 3.4. The number of watts radiated by this source may be obtained from Planck's equation particularized to 2042°C and $1/60$ cm^2 and integrated over all energies (wavelengths).

Table 3.2.   Sources of Background Illumination

Office—large retail outlets	Diffuse fluorescent
	25–100 fL
	50 fL average
Laboratory	Diffuse
	Point—diffuse
	Windows, lamp
Small retail outlets	Fluorescent
	Windows
Auto, aircraft	Indirect sunlight
Night vision	Starlight, moonlight, sky scatter

Light source characteristics

Point source (approximate)	
Coherent	Laser
Noncoherent	Incandescent bulb
	Sun
Line source	Fluorescent
Diffuse source	Lamps with shades or diffusers
	"Bounced" sources

$$W_\lambda = \frac{2\pi c^2 h}{\lambda^5[\exp(hc/\lambda kT) - 1]} \qquad \begin{array}{l} \text{W/m}^2 \text{ per unit wavelength} \\ \text{Planck's equation} \end{array}$$

where λ = wavelength

h = Planck's constant (6.626×10^{-34} W)

c = velocity of light in free space (2.998×10^8 m/s)

k = Boltzmann's constant (1.380×10^{-23} J/K)

T = absolute temperature, K

and the radiant exitance is

$$W = \int_0^\infty W_\lambda \, d\lambda = \sigma T^4 \qquad \text{W/m}^2$$

where the Stefan-Boltzmann constant σ is:

$$\sigma = \frac{2\pi^5 k^4}{15c^2 h^3} = 5.670 \times 10^{-8} \text{ W} \cdot \text{m}^{-2} \cdot \text{K}^{-4}$$

The source power is

$$9.916 \times 10^5 \times 1/60 \times (10^{-2})^2 = 1.643 \text{ W}$$

for the ($\frac{1}{60}$-cm^2 area) 1-cd source.

From Fig. 3.2 note that $M = \pi L$ for a Lambertian source such as a blackbody and has constant luminance L independent of direction of view. Its luminous exitance based on a 60-cd/cm^2 luminance, L, is

$$M = \pi L = 188.5 \text{ lm/cm}^2$$

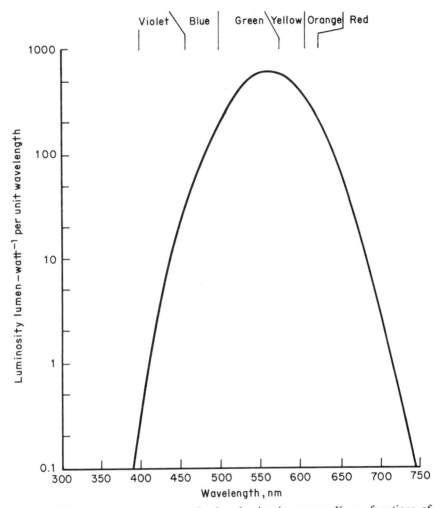

Fig. 3.3. CIE eye response curve. Absolute luminosity curves K_λ as functions of wavelength (response of the human eye to radiation on a given wavelength).

Its luminous efficacy K is

$$K = \frac{M}{W} = 1.91 \text{ lm/W}$$

Alternatively, the definition of the candela could be verified by combining the eye response and blackbody curves:

$$L = \int_0^\infty K\lambda I_\lambda \, d\lambda \qquad \text{cd/m}^2$$

with the result

$$L = 60 \times 10^4 \text{ cd/m}^2$$

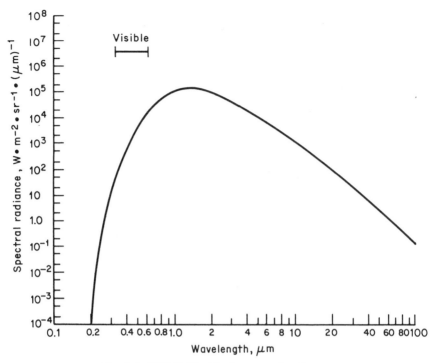

Fig. 3.4. 2042 K Blackbody spectral radiance.

The brightness of an emitter which radiates in a narrow spectral band such as Fig. 3.5 may be calculated in a similar fashion as

$$L = \int_0^\infty K_\lambda \frac{W_\lambda}{\pi} \, d\lambda$$

where use of $I_\lambda = W_\lambda/\pi$ makes the assumption that the source is Lambertian, and W_λ now is power per unit energy interval of the light source.

For the case where the display spectrum is given as relative power output, and total power output is also specified or measured, L may be determined by consideration of these two quantities as follows:

$$W_0 = A \int_0^\infty W_{R\lambda} \, d\lambda$$

where W_0 = measured total power
$W_{R\lambda}$ = reported spectrum
A = normalizing factor

The integral may be calculated by counting squares.

L is then calculated as

$$L = \frac{W_0}{\displaystyle\int_0^\infty W_{R\lambda} \, d\lambda} \times \int_0^\infty K_\lambda W_{R\lambda} \, d\lambda$$

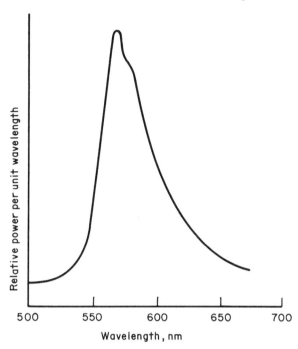

Fig. 3.5. Spectrum of a GaP:N VLED.

Various photometric quantities are summarized in Table 3.3. Some useful conversion factors are given in Table 3.4. Some definitions are also given in Appendix 3.1.

Photometric measurements are made by use of a photometer (Fig. 3.6). Commercial photometers generally image the source to be measured on a silicon detector. The detector is filtered to approximate the CIE curve of Fig. 3.3. Large deviations can occur in the red where the silicon is highly sensitive and the eye is not, so special correction factors for use with specific sources such as red LEDs are commonly provided. Secondary standard sources are also available for calibration of these photometers from the same manufacturers.

Photometers are usually calibrated in footlamberts or lux. Such a calibration is based on the assumptions:

1. The source is uniform inside and outside the measurement spot; i.e., a segment brightness is uniform and a measurement at the center is representative. This restriction may be overcome by scanning the segment vertically and horizontally and normalizing the scan to the center reading.
2. The source is Lambertian; i.e., its brightness per unit solid angle has a cosine viewing angle dependence from the normal above the half plane. This assumption allows luminous flux at a detector normal to an emitter to be converted from lumens to footlamberts.

Assumption 2 is not particularly critical since most measurements are made normal to the segment or dot, and usually used for display-to-display intercomparisons, or for design purposes in conjunction with an angular response plot.

Table 3.3a. **Symbols, Defining Equations, and Standard Units for Fundamental Photometric Quantities**

Quantity*	Symbol*	Defining equation†	Commonly used units	Symbol
Luminous energy (quantity of light)	$Q_e(Q_v)$	$Q_v = \int_{380}^{780} K(\lambda)Q_{e\lambda}\,d\lambda$	lumen-second† (talbot)	lm· s
Luminous density	$w_e(w_v)$	$w = dQ/dV$	lumen-second per cubic meter‡	lm· s· m^{-3}
Luminous flux	$\Phi_e(\Phi_v)$	$\Phi = dQ/dt$	lumen‡	lm
Luminous exitance	$M_e\ (M_v)$	$M = d\Phi/dA$	lux‡ (lm/m²)	lx
Luminous intensity (candlepower)	I_e/I_v	$I = d\Phi/d\omega$ (ω = solid angle through which flux from point source is radiated)	candela‡ (lumen per steradian)	cd
Luminance (photometric brightness)	$L_e(L_v)$	$L = d^2\Phi/d\omega(dA\cos\theta)$ $= dI/(dA\cos\theta)$ (θ = angle between line of sight and normal to surface considered)	nit (cd/m²‡)	nt
Luminous efficacy	K	$K = \Phi_v/\Phi_c$	lumen per watt†	lm/W
Luminous efficiency, spectral	$V(\lambda)$	$V(\lambda) = K(\lambda)/K(\lambda)_{max}$	numeric	

*Quantities may be restricted to a narrow wavelength band by adding the word spectral and indicating the wavelength. The corresponding symbols are changed by adding a subscript λ, for example, Q_λ for a spectral concentration or a λ in parentheses, for example, $K(\lambda)$, for a function of wavelength.

†The equations in this column are given merely for identification.

‡International System (SI) unit.

NOTE: The symbols for radiometric quantities (see next page) are the same as those for the corresponding photometric quantities (see above). When it is necessary to differentiate them, the subscripts e and v respectively should be used, for example, Q_e and Q_v.

SOURCE: Ref. 10.

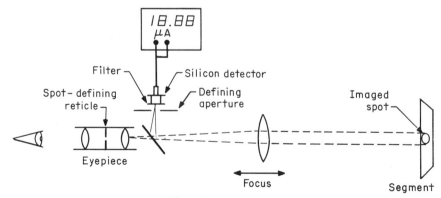

Fig. 3.6. Microphotometer construction.

Table 3.3b. Symbols, Defining Equations, and Standard Units for Fundamental Radiometric Quantities

Quantity*	Symbol*	Defining equation†	Commonly used units‡	Symbol
Radiant energy	$Q_v(Q_e)$		joule	J
Radiant density	$w_v(w_e)$	$w = dQ/dV$	joule per cubic meter	J/m^3
Radiant flux	$\Phi_v(\Phi_e)$	$\Phi = dQ/dt$	watt	W
Radiant exitance	$M_v(M_e)$	$M = d\Phi/dA$	watt per square meter	W/m^2
Radiant intensity	$I_v(I_e)$	$I = d\Phi/d\omega$ (ω = solid angle through which flux from point source is radiated)	watt per steradian	W/sr
Radiance	$L_v(L_e)$	$L = d^2\Phi/d\omega\,(dA \cos\theta)$ $= dI/(dA \cos\theta)$ (θ = angle between line of sight and normal to surface considered)	watt per steradian square meter	$W \cdot sr^{-1} \cdot m^{-2}$
Emissivity	ϵ	$\epsilon = M/M_{bb}$ (M and M_{bb} are respectively the radiant exitance of the measured specimen and that of a blackbody at the same temperature as the specimen)	numeric	

*Quantities may be restricted to a narrow wavelength band by adding the word spectral and indicating the wavelength. The corresponding symbols are changed by adding a subscript λ, for example, Q_λ for a spectral concentration or a λ in parentheses for example, $K(\lambda)$, for a function of wavelength.

†The equations in this column are given merely for identification.

‡Only International System (SI) units are shown.

SOURCE: Ref. 10.

A point source is one which at standard viewing distance of 0.46 m (18 in) subtends an arc less than that resolved by the eye of about 40 seconds under bright illumination, or about 10 micrometers (Figs. 3.7 and 3.8). Such a source can be measured by use of an integrating sphere which is a detector which collects all the radiation. Some types of integrating spheres are shown in Fig. 3.9. They can be calibrated with a standard source provided only sources with the same specification are measured.

If the source spectra and detector spectral response have previously been determined, either from a data sheet or by use of a spectrometer, the calibration factor is calculated as described earlier for the case of a narrow-band emitter and eye detector. A standard source is often used to provide the power calibration.

Contrast Enhancement

The display background is a direct function of the illuminance of the surroundings or ''surround.'' Typical surround brightnesses are given in Table 3.4. Figure 3.10

Table 3.4a.　Useful Conversion Factors for Optoelectronics

To convert from		To		Multiply by
		Length and area		
angstrom	(Å)	meter	(m)	$1.000\ 000 \times 10^{-10*}$
angstrom	(Å)	micrometer	(μm)	$1.000\ 000 \times 10^{-4*}$
foot	(ft)	meter	(m)	$3.048\ 000 \times 10^{-1*}$
square foot	(ft^2)	square meter	(m^2)	$9.290\ 304 \times 10^{-2*}$
inch	(in)	meter	(m)	$2.540\ 000 \times 10^{-2*}$
micron	(μ)	micrometer	(μm)	$1.000\ 000*$
		Photometry		
apostilb	(asb)	nit	(nt)	$3.183\ 099 \times 10^{-1}$
candela	(cd)	lumen per steradian	(lm/sr)	$1.000\ 000*$
candela per square foot	(cd/ft^2)	nit	(nt)	$1.076\ 391 \times 10$
footcandle (lm/ft^2)	(fc)	lux	(lx)	$1.076\ 391 \times 10$
footlambert	(fL)	candela per square foot	(cd/ft^2)	$3.183\ 099 \times 10^{-1}$
footlambert	(fL)	nit	(nt)	$3.426\ 259$
lambert	(L)	nit	(nt)	$3.183\ 099 \times 10^{3}$
lux	(lx)	lumen per square meter	(lm/m^2)	$1.000\ 000*$
nit	(nt)	candela per square meter	(cd/m^2)	$1.000\ 000*$
phot (lm/cm^2)	ph	lux	(lx)	$1.000\ 000 \times 10^{4*}$
stilb (cd/cm^2)	(sb)	nit	(nt)	$1.000\ 000 \times 10^{4*}$

*The asterisk indicates an exact conversion. All others are rounded.
SOURCE: Ref. 10.

Table 3.4b.　Luminance Conversion Factors
(Read across and up)

	Candela/cm²	Candela/m²	Equivalent phot	mL	Equivalent lux	Equivalent fc (fL)	Candela/ft²
1 candela/cm² (stilb) =	1	10^4	π	$\pi \cdot 10^3$	$\pi \cdot 10^4$	2.92×10^3	9.29×10^2
1 candela/m² (nit) =	10^{-4}	1	$\pi \cdot 10^{-4}$	$\pi \cdot 10^{-1}$	π	0.292	9.29×10^{-2}
1 equivalent phot (L) =	0.3183	3.183×10^3	1	10^3	10^4	9.29×10^2	2.96×10^2
1 mL =	3.183×10^{-4}	3.183	10^{-3}	1	10	0.929	0.296
1 equivalent lux (blondel) (apostilb) =	3.183×10^{-5}	0.3183	10^{-4}	10^{-1}	1	9.29×10^{-2}	2.96×10^{-2}
1 equivalent fc (fL) =	3.426×10^{-4}	3.426	1.076×10^{-3}	1.076	10.76	1	0.3183
1 candela/ft² =	1.076×10^{-3}	10.76	3.382×10^{-3}	3.382	33.82	π	1

SOURCE: Ref. 11.

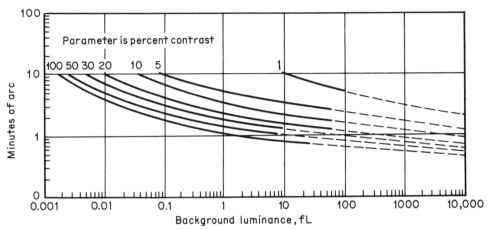

Fig. 3.7. Visual acuity as a function of contrast. *(After Ref. 1, as adapted from Carel.)*

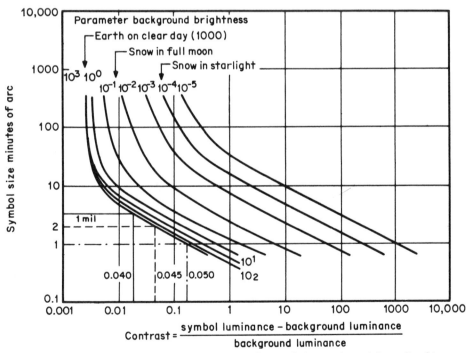

$$\text{Contrast} = \frac{\text{symbol luminance} - \text{background luminance}}{\text{background luminance}}$$

Fig. 3.8. Contrast thresholds of the eye. *(After Ref. 1, as adapted from Carel.)*

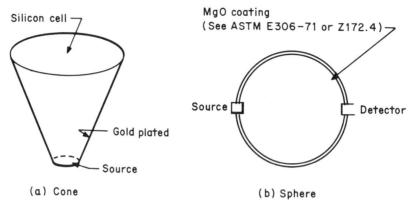

Silicon cell

MgO coating
(See ASTM E306-71 or Z172.4)

Source

Detector

Gold plated

Source

(a) Cone

(b) Sphere

Fig. 3.9. Representative integrating spheres and cones. They may be calibrated with a standard source, provided the unknown source has the same emission spectra as the standard.

shows radiation from the surround incident on the display through a filter display cover as discussed in Chap. 1. A number of cases may be considered with reference to this figure. For example, if a point source of light is incident on a flat smooth cover at an angle θ greater than the critical angle (42° for glass or plastic), the cover will act like a mirror and the viewer will see the source totally reflected at θ, or the other side of the normal. For the TI-150 with about a 6-cm-wide filter and a lamp incident at 30° over the left shoulder, the resulting angle to the normal can be as much as 60° to the right eye, although the bulb is shielded from the right eye. If the head moves from a 30° angle between the normal and the right eye at the left side of the display to a 30° angle at the right side, the bulb image moves to the right so that the left eye will start to see the bulb on the right side, and so on as the head continues to move. Note the TI-150 bezel is recessed about $\frac{1}{2}$ to 1 cm, which mitigates glare from such sources.

In referring to the figures, the sources of lit-segment contrast reduction are: (1) unlit segment highlighted by lit segment, (2) display background highlighted by lit segment, along with (3) contributions from incoming illumination, (a) reflection from lit segment, (b) reflection from unlit segment, (c) reflection from cover, and (d) reflection from background.

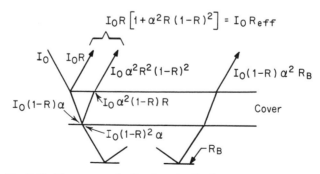

$$I_0 R \left[1 + \alpha^2 R (1-R)^2 \right] = I_0 R_{eff}$$

I_0 $I_0 R$

$I_0 \alpha^2 R^2 (1-R)^2$ $I_0 (1-R) \alpha^2 R_B$

$I_0 (1-R) \alpha$ $I_0 \alpha^2 (1-R) R$ Cover

$I_0 (1-R)^2 \alpha$

R_B

Fig. 3.10. Geometry of reflections in display cover and from display segments and background.

The first case considered earlier may be extended to that of a gray cover with front surface glare:

RI_B where R is the cover reflectivity and I_B is the surround illumination. The veiling glare I_V is

$$I_V = I_B R_{\text{eff}}$$

where R_{eff} is defined in Fig. 3.10.

Let $R = 0.07$ and $\alpha = 0.4$; then R_{eff} is 0.08. The segment will not be seen through the cover when $I_0 \times 0.08 \geq (1 - 0.08) I_{\text{seg}}$, or when I_0 is about 10 times brighter than the segment.

For Case 2, it is seen that the choice of the background parameter removes it from view; i.e., a gray or black background (R_B less than $\frac{2}{3}$ for the above example) cannot be seen through the cover glass.

In certain cases shiny segments will be visible as illustrated in Fig. 3.10.

Contrast enhancement depends on the control of reflections from three regions: cover, background, lit segment. It can be seen that cover glass reflections can be reduced by: (1) changing it from a smooth specular or mirror surface to a curved and pebbled or diffuse surface. (Care must be taken to avoid making the surface opaque and reducing the crispness of display.) Antireflective coatings are also effective; (2) eliminating scratches and moving edges away from the field of view to avoid concentrating veiling.

Backgrounds must be flat black or gray. Shiny leads and segments must be avoided or painted black.

In addition, there are contrast enhancement techniques. Most of these reduce the viewing angle or transmitted light (Ref. 1, p. 461).

- Hoods as provided by recessed bezels
- Shuttered covers, such as gridded micromesh or honeycomb
- Cross mesh or wire
- Circular polarizer
- Neutral filter and thin film

Neutral-Density Filter. The neutral-density filter reduces the terms of Case 1 by introducing an absorption term which is doubled for the surround illumination and occurs only once for the lit segment. The neutral-density filter absorbs light of all colors equally. If a filter with 40% transmissivity is used (where the 40% can include about 50% absorption and 10% first- and second-surface reflection), then 40% of the display output is transmitted, and at most, 16% of the background. Where the background has 10% reflectance and the surround is twice as bright as the display, the contrast is enhanced from

$$\frac{1 - 0.20}{0.2} = 40$$

to

$$\frac{40\% - 3.2\%}{3.2\%} = 115$$

Neutral-density filters are available in glass and acrylic plastics.

Circular polarizers operate on the principle that specularly reflected light is polarized; therefore, polarization of the surround illuminance by specular reflection from a display cover will be attenuated on the second pass through the polarizer. The polarizer is most effective for glass packaged displays such as gas discharge and vacuum fluorescent displays. However, polarizers suffer from the disadvantages of high cost. Often curved glass display surfaces are used, which can reduce specular reflections to a small spot or line reflection.

Application information on louver films may be obtained from the 3M Company. Bandpass filters will be discussed as an aid to color contrast enhancement.

Color Contrast

Display contrast can be provided by color as well as brightness; however, it can be a hindrance as well as an aid. For example, segment-to-segment color differences can be objectionable. Just discernible color contrast is also a function of color (wavelength) (Fig. 3.11). For our purposes, the color names will be associated with wavelength regions as given in Table 3.5. It can be seen that (1) there is some difference on color names by color boundary between experimental groups, and (2) some colors have very narrow boundaries. This result is also reflected in Fig. 3.11. Hue and saturation (Refs. 12 and 13) will not be important to the display designer

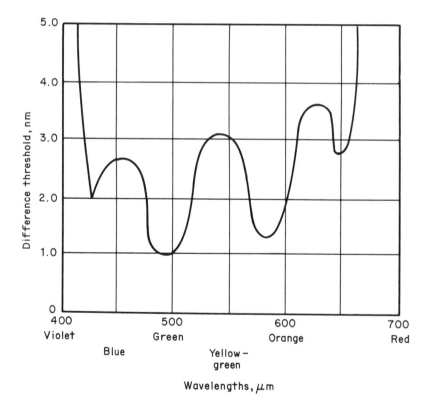

Fig. 3.11. Smallest difference in wavelength that can be detected as a difference in hue by using the comparative method of discrimination. *(After Ref. 14.)*

Table 3.5. Color Names

Color names after Bergh and Dean, based on OSA[12]	
Red	700 nm
Yellow	577 nm (equal parts red and green)
Green	546.1 nm
Blue	435.8 nm

Color names after Loebner, based on CIE[13]	
Red	660 ± 30 nm
Yellow	575 ± 5 nm
Green	510 ± 30 nm
Blue	470 ± 10 nm

because he or she will have little control over them, and because they do not enter into design equations.

Typically, color contrast is treated by filtering the display and treating the case as one of brightness contrast.

Examples of red and green VLEDs (visible light-emitting displays) in fluorescent light and sunlight are treated by Pucilowski, Schuman, and Velasquez[2] and are shown in Table 3.6, where the appropriate contrast formulas from Table 3.1 are used for the calculation. Some sources, filters, and VLED spectra used in the calculation are shown in Figs. 3.12, 3.13, and 3.14. It can be seen that green has low contrast in the ambients considered. Also compared to the colored-filter performance is neutral-density-filter performance (see the table).

Evans[3] has also considered the use of colored filters to enhance brightness contrast, although he presents no contrast calculations. However, a curve for the bandpass filter for green (Fig. 3.15) can be compared with the curve for yellow (Fig. 3.16).

Filters. Plastic filters, as described above, are generally not available in plastic sheet form. The usual practice is to obtain dye lot samples in the form of small stepped thickness sheets. Eyeball tests then complement designs based on the above considerations to select transmission (sheet thickness) and color. Then an

Table 3.6. Contrast Examples: Fluorescent Light Ambient

Ambient illumination	Red or green LED (Filter 1)	Green LED (Filter 4)	Red LED (Filter 7)
100 lx	12.8	11.7	829
300 lx	4.93	4.58	277
500 lx (normal room lighting)	3.36	3.15	167
1000 lx	2.18	2.07	84
LED luminance out (nt)	60	111	119

Filter 1: neutral density filter $T_n = 0.4$.
Filter 4: ideal bandpass filter (Fig. 3.13).
Filter 7: ideal bandpass filter (Fig. 3.14).
SOURCE: Adapted from Ref. 2.

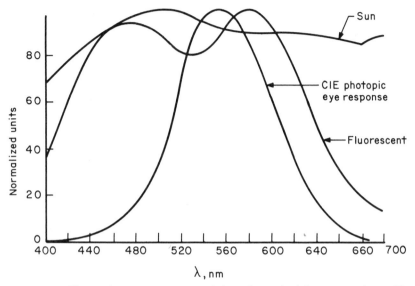

Fig. 3.12. The sun's power spectra and that of a typical fluorescent lamp. The CIE photopic eye response is also shown. *(After Ref. 2.)*

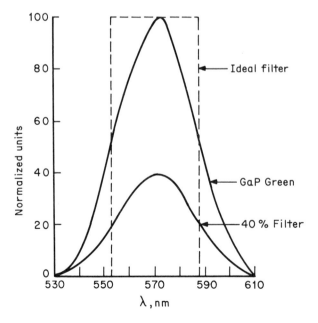

Fig. 3.13. Green LED (GaP:N) spectra compared to the transmission of an ideal green filter 4 and an ideal bandpass filter. *(After Ref. 2.)*

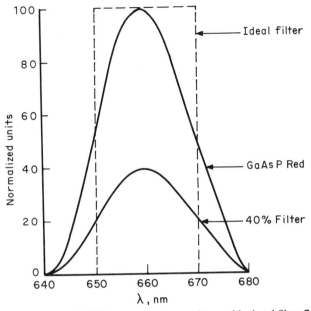

Fig. 3.14. A red LED spectra compared to an ideal red filter 7 and a hypothetical bandpass filter. *(After Ref. 2.)*

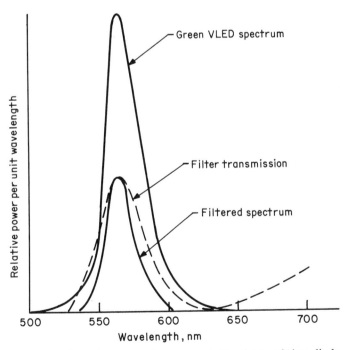

Fig. 3.15. Filtered output of a green visible light-emitting diode (VLED). The output is reduced in the example by $\frac{2}{3}$. *(Adapted from Ref. 3.)*

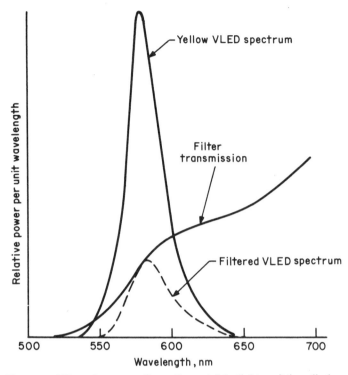

Fig. 3.16. Filtered output of a yellow visible light-emitting diode (VLED). The output in the example is reduced by ¾. *(Adapted from Ref. 3.)*

independent plastic manufacturer is obtained to form the molding powder of the selected dye lots into suitable shapes.

Finally, Ellis, Burrell, Wharf, and Hawkins[4] found that the luminance of a green dot-matrix display must exceed that of a red one by a factor of 3 when viewed under 10^5 lx from a quartz halogen lamp (tungsten filament) for equal legibility, and also with the contrast for the tungsten ambient, which is seen to reduce the contrast for the red.

Human Factors: Minimum Contrast

Although it is desirable to separate contrast from size effect requirements, this is not possible, as shown in Fig. 3.7. While size cannot be considered separately from minimum contrast requirements, the size will apparently vary by no more than 4 times over a contrast range from 1 to 10. Since it will be seen later that minimum contrast requirements will need to be multiplied by 3 to 4 times to obtain comfortable viewing, it will be argued here that—to zero order—size may be neglected when specifying contrast. Iterative procedures, not now available, can allow improvements in estimates.

Direction of contrast is a consideration. LCDs and other passive displays generally use negative contrast or dark symbols on light background. The authors or Ref. 1 (p. 471) apparently conclude that symbol of visibility is not much affected by contrast polarity, whereas legibility may be. Color contrast human factor recommendations are (Ref. 1, p. 475):

The reader must be aware, therefore, that symbol-background color contrast is a significant design variable and that he must exercise his own variety of caution in specifying legibility contrast ratios for symbology or other than white or green-yellow color viewed against gray backgrounds since it is these wavelengths and wavelength combinations upon which the majority of design-oriented contrast ratio data have been taken.

Many estimates of required contrast are based on Blackwell's 1946 study (Ref. 5), where contrast is shown to be a function of the background illumination. Carel[6] has adapted his data to aid contrast calculations (Fig. 3.8). These contrast ratios are for visibility 50% of the time; multiply by 3 to estimate 99%. Figure 3.7 relates visual acuity requirements to background and contrast. Again multiply by 3 and use the highest value obtained from the two curves.

Carel reports that the use of contrasts calculated from the figures leads to ghostly images and recommends multiplying by a factor of 5 (Muller's factor).

An example from Ref. 1 serves to illustrate the procedure. Suppose display segments as viewed at 0.46 m (18 in) subtend 3.4 minutes of arc at the eye; the display background luminance is 20 fL and the surround luminance (sky or room) is 200 fL. Then from 3.8 the 50% contrast threshold for a 3.4-minute symbol against a 20-fL background will be 0.03 (or 3%). A value from Fig. 3.7 of 0.05 is obtained; therefore, 0.05 is used in subsequent estimates. Multiply by 3 to correct to 99% and by 5 to eliminate the ghostly image problem; the result is 0.75. The background and surround luminance differ by 10 times and an eye adaptation mismatch may be applied from Fig. 3.17 of 1.2, raising the contrast ratio to .90.

Since $E/B = 0.9$ and B is 20 fL, E is 18 fL more than B (through the filter) and E is 38 fL. This result may be compared to another study by King (Fig. 3.18) where a

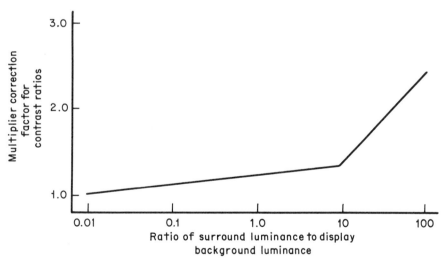

Fig. 3.17. Correction factors for eye adaptation mismatch. *After Ref. 1, as adapted from Carel, Ref. 6.)*

contrast of .95 is suggested for numerics. Note that the numerics curve differs from the bar curve by a factor of about 4.

Finally, a comfort level factor is suggested in Reference 1 (p. 493), varying from high values at low backgrounds as 4 to 3 for bars, and 2.5 to 1.1 for numerics (B of 10 fL and 200 fL, respectively). As applied to the preceding example, the display through the filter should be about 80 fL. Based on commercial calculator practice where values of 30 to 100 fL are common for use in 50 to 100 fL surrounds, these values seem quite reasonable.

Summary

In summary, calculations are a guide to design. They can show required and expected improvements and performance penalties. Second, variations in surrounds lead to severe design problems. Third, there is no substitute for using and looking.

3.3 LEGIBILITY

The definition of legibility has been given by McCormic[7] as:

the attribute of being able to identify given letters or numerals to the exclusion of others depends primarily on such features as stroke width, form of character, background, size and illumination.

In Ref. 1, legibility is defined as "a property of alphanumerics which is measured in terms of three objective performance criteria: accuracy, speed, and rate of symbol identification."

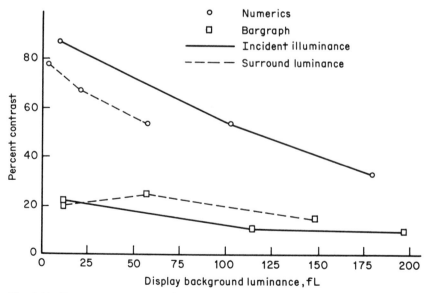

Fig. 3.18. Percent contrast required for 99% display legibility. *(After Ref. 1, as adapted from King, Ref. 15.)*

The principal tools of legibility research are the confusion matrix, identification times, and accuracy of identification where the confusion matrix is a frequency of misidentification by type matrix.

Components of legibility include (in addition to contrast, brightness, and flicker): font, size, width-to-height ratio, stroke-to-height ratio, spacing, viewing angle (foreshortening), sharpness (acutance, blur, granularity), jitter, smear (lag).

Insofar as human factors yield design information to guide display design, most of its results are font dependent. However, font dependency will be discounted by making the assumption that with enough practice, "well-designed" fonts are equally readable. (Note, however, that this assumption tends to minimize the effect of nearly all the variables above on legibility.) For font studies see, for example, Refs. 1 and 4.

Character Size and Proportion

Height. Human factors research leads to minimum viewing-angle requirements which is to the required display height by:

$$\tan \alpha = \frac{H}{D}$$

where H = character height

D = distance from the eye to the display in the same units as H

and for the small angles of interest:

$$\alpha = 3600 \frac{H}{D}$$

where the result is in units of minutes.

Reference 1 summarizes the relevant research as at least 15 minutes of arc is required for symbology on the CRT screen. Duncan and Konz[8] recommend 20 minutes of arc for the VLED display, which for 0.6-m viewing distance indicates a 3.5-mm height (5.8 mm at 1 m). For the transmissive liquid crystal display, they recommend 26 minutes of arc, and for the reflective LCD about 30 minutes of arc when the illumination is in the range 16 to 484 lx. They also note that LEDs may be readable at 4 minutes of arc, transmissive LCDs at 5 minutes of arc, and reflective LCDs at 6 minutes of arc with normal vision.

Symbol Width to Height and Spacing. The authors of Reference 1 recommend W/H ratios in the range of ½ to 1. They recommend segment width (stroke width) to height ratios in the range of 13 to 20%. They recommend spacing-to-height ratios in the range of 26 to 63%.

They also recommend that viewing angles be held to 19° from the normal.

Sharpness and Blur

Fuzzy or jittery edges tend to wash out narrow segments or small dots, such as on gas discharge displays, but don't appreciably affect the appearance of broad segments, such as on vacuum fluorescent displays. Sharpness or acutance may be defined in terms of knife-edge parameters or step-wedge MTF (modulation transfer function)[9] and measured by scanning across a segment with a spot photometer and deconvolving the spot blurring. However, neither calculation nor measurement are commonly done since blur may be offset by increased contrast[1] and is often used to increase apparent brightness for marginal manufacturing processes. Ragged (grainy) segments apparently are not objectionable as long as the segment notches don't exceed about one-fourth the width. Reference 1 notes that both flicker and graininess degrade legibility.

Summary

Legibility recommendations are summarized in Table 3.7.

3.4 TEMPORAL FACTORS

Factors that depend on time are viewer fatigue, data entry time, and flicker. There appears to be little data on viewer fatigue, but user use for more than 4 hours continuous operation would indicate that comfortable contrast, color, and legibility, as recommended earlier, and comfortable surround illumination, are all important.

Digits or characters may be slow to enter, but speeds greater than 0.1 to 0.3 s apparently are acceptable.

Flicker affects the viewer as shown in Fig. 3.19. Flicker is not perceived above a rate called the *critical fusion frequency* (cff). The cff varies from observer to observer and depends on the factors shown in Table 3.8 from Ref. 1. However, use

Table 3.7. Summary of Legibility Requirements

Factor	Recommendations
Height	20 minutes* for VLED 26 minutes for transmissive LCD 30 minutes for reflective LCD
Symbol:	
Width to height	1/2 to 1
Segment width to height	13 to 20%
Spacing to height	26 to 63%
Viewing angles (to normal):	0 to 19°
Sharpness	Increase the brightness for fuzzy character displays over that required for sharp.
Contrast	Discussed in earlier section
Not discussed[1]	
Display background luminance	Filter types
Eye adaptation levels	Emitter reflectance
Symbol type or font	Color
Contrast polarity	Variable interactions
Display vibration	

*Minutes of arc.

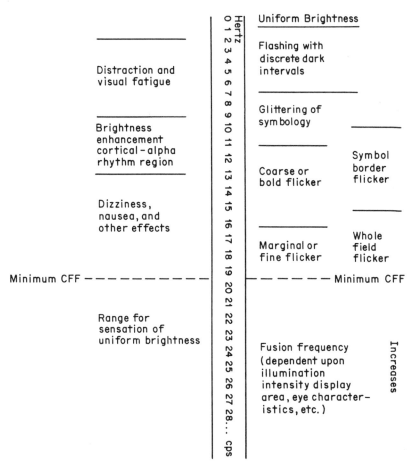

Fig. 3.19. Approximate flicker stages and resulting observer reactions. *(After Ref. 1, p. 438.)*

of refresh rates exceeding 60 Hz will exceed most individuals' cff in most circumstances (one exception will be rapid relative motion between display and eye).

REFERENCES

1. C. A. Semple, Jr., R. J. Heapy, E. J. Conway, Jr., and K. T. Burnette, *Analysis of Human Factors Data for Electronic Flight Display Systems,* Technical Report AFFDL-TR-70-174 (1971). Available as AD884770 from U.S. Department of Commerce.
2. J. Pucilowski, R. Schuman, and J. Velasquez, *Appl. Opt.,* **13** (1974), p. 2248.
3. D. Evans, *Electron. Des.,* **17** (1976), p. 62.
4. B. Ellis, G. J. Burrell, J. H. Wharf, and T. D. F. Hawkins, *Proc. of the Soc. for Inf. Disp.,* **16** (1975), p. 250.
5. H. R. Blackwell, *J. Opt. Soc. Am.,* **36** (1946), p. 624.

Table 3.8. Representative Observer-Display Interactions

Observer parameters	Display parameters											
	Phosphor persistence rate	Emitted luminescence intensity	Emitted spectral quality	Regeneration rate	Update rate	Viewing distance	Screen size	Contrast	Light-dark ratio	Display orientation	Surround illumination	Writing speed
Persistence of vision	×	×	×	×			×		×		×	×
Foveal-cortex character		×	×	×	×		×		×		×	
Chromatic aberration		×	×			×	×	×		×	×	
Spherical aberration		×	×			×				×		
Foveal area simulated	×	×	×	×	×	×	×	×	×	×	×	×
Eye adaptation level	×	×	×	×			×	×	×		×	×
Spectral sensitivity range		×	×	×				×	×		×	
Retinal illumination	×	×	×	×	×	×	×	×	×	×	×	
Viewing angle		×	×			×	×	×		×	×	
Brightness enhancement	×	×	×	×			×	×	×		×	
Pupil size							×	×		×		
Critical fusion frequency	×	×	×	×	×	×	×	×	×	×	×	×

NOTE: The existing body of flicker data is large and does provide sufficient basis for the making of firm decisions, provided caution is exercised in applications to new situations. It is estimated that flicker-free CRT displays can be designed using the commonly accepted 50 to 60 Hz refresh rate, even under relatively high illumination conditions. However, displays designed for use under extremely high luminance environments (8000 fL and up) would probably require additional validation, if not additional research and experimentation.

SOURCE: Ref. 1.

54

6. W. L. Carel, *Pictorial Displays for Flight,* Office of Naval Research (1965). Available as AD637669 from U.S. Department of Commerce.
7. E. J. McCormic, *Human Factors Engineering,* McGraw-Hill, New York (1970).
8. J. Duncan and S. Konz, *Proc. of the Soc. for Inf. Disp.,* **17** (1976), p. 180.
9. H. R. Luxenberg and R. L. Kuehn (eds.), *Display Systems Engineering,* McGraw-Hill, New York (1968), p. 153.
10. From USAS Z7.1-1967, Nomenclature and Definitions for Illuminating Engineering, as abstracted in JEDEC publication no. 77A.
11. R. H. Akin and J. M. Hood, "Photometry," in H. R. Luxenberg and R. L. Keuhn (eds.), *Display Systems Engineering,* McGraw-Hill, New York (1968), Chapter 3.
12. Optical Society of America, Committee on Colorimetry, *The Science of Color,* Crowell, New York (1953), as reported in: A. A. Bergh and P. J. Dean, *Light-Emitting Diodes,* Clarendon Press, Oxford (1976), p. 50.
13. Publ. CIE2N.I.3.3 Paris, France, 1959, as reported in: E. E. Loebner, *Proc. IEEE,* **61** (1973), p. 837.
14. J. W. Wulfeck, A. Weisz, and M. Raben, *Vision in Military Aviation.* Air Res. and Dev. Coms. WADC, WPAFB, Ohio (1958), WADC-TR-58-399 (AD207780).
15. R. C. King, R. W. Wollentin, C. A. Semple, and G. Goettelman, *Electroluminescent Display Legibility Research and Development,* Air Force Systems Command, WPAFB, Ohio (1970), AFFDL-TR-70-89, from Ref. 1.

APPENDIX 3.1
Photometric Units: A Glossary of Terms*

Build-up (of luminance)   The increase in luminance with time, under repeated excitation, from the initiation of excitation to the point where equilibrium occurs or excitation ceases.

Color   Color consists of the characteristics of light, other than brightness or luminance, by which a human observer may distinguish between two structure-free patches of light of the same size and shape.

Note: Also refer to JEDEC Publication No. 16C for specification or description of color.

Flicker   The sensation of brightness or color variation, occurring when the frequency of the observed variation is between a few hertz and the flicker fusion frequency of the image.

Flicker fusion frequency (FFF)   The frequency of intermittent stimulation of the eye at which flicker disappears. It is also called the critical fusion frequency (CFF) or the critical flicker frequency (CFF).

Footlambert (fL)   A unit of luminance (photometric brightness) equal to $1/\pi$ candela per square foot, or to the uniform luminance of a perfectly diffusing surface emitting or reflecting light at the rate of one lumen per square foot, or to the average luminance of any surface emitting or reflecting light at that rate.

Note 1: The average luminance of any reflecting surface in footlamberts is, therefore, the product of the illumination in footcandles by the luminous reflectance of the surface.

Note 2: The preferred unit of luminance is the nit. Conversion: nit = $3.4258 \times$ fL.

Glare   The sensation produced by luminances from extraneous sources within the visual field that are sufficiently greater than the luminance to which the eyes are adapted to cause annoyance, discomfort, or loss in visual performance and visibility.

Note: The magnitude of the sensation of glare depends upon such factors as the size,

*From *SID 16/4,* 1975, p. 259, and jointly with EIA as JEDEC No. 92, January 1975, with permission.

position, and luminance of a source, the number of sources, and the luminance to which the eyes are adapted.

Halation   The presence of an illuminated annular area surrounding the spot, caused by internal reflection from the front and rear surfaces of the faceplate, of luminance generated by the electron beam.

Luminance (photometric brightness), $L = d^2\Phi/d\omega(dA \cos \theta)$ Luminance (photometric brightness) in a direction, at a point of the surface of a source, of a receiver, or of any other real or virtual surface is the quotient of the luminous flux (Φ) leaving, passing through, or arriving at an element of the surface surrounding the point, and propagated in directions defined by an elementary cone containing the given direction, by the product of the solid angle ($d\omega$) of the cone and the area (dA) of the orthogonal projection of the element of the surface on a plane perpendicular to the given direction; or it is the luminous intensity of any surface in a given direction per unit of projected area of the surface as viewed from that direction. A single unit of luminance, the nit, should be used and all others decried. The footlambert should be included as a temporary measure during the change of nomenclature. The test should specify whether a spot, line, or raster is used as well as other test parameters.

Note: In the defining equation θ is the angle between the direction of observation and the normal to the surface. In common usage the term *brightness* usually refers to the intensity of sensation which results from viewing surfaces or spaces from which light comes to the eye. This sensation is determined in part by the definitely measurable luminance (photometric brightness) defined above, and in part by conditions of observation such as the state of adaptation of the eye. (See **subjective brightness**.) In much of the literature the term *brightness* used alone refers to both luminance and sensation. The context usually indicates which meaning is intended.

Luminance build-up   The increase in luminance with time, under repeated excitation, from the initiation of excitation to the point where equilibrium occurs or excitation ceases.

Luminance contrast ratio   The ratio of the total luminance at any information element to the background or surround luminance. It is given by

$$C_R = \frac{L_1}{L_2}$$

where L_1 = Total luminance at information element

L_2 = total background or surround luminance

Note: If the background luminance is greater than the luminance at the information element, then L_1 and L_2 should be interchanged.

Nit (nt)   The unit of luminance (photometric brightness) equal to one candela per square meter. (Conversion: nit × 0.2919 = fL.)

Note: Candela per square meter is the International System (SI) unit of luminance.

Persistence   The continuation of luminance or radiance after the excitation has been removed.

Persistence characteristic (decay characteristic)   The relation, usually shown by a graph, between the luminance and the time elapsed after the removal of the excitation.

Radiance, $L = d^2\Phi/d\omega(dA \cos \theta)$ Radiance in a direction, at a point of the surface of a source, of a receiver, or of any other real or virtual surface is the quotient of the radiant flux (Φ) leaving, passing through, or arriving at an element (dA) of the surface surrounding the point, and propagated in directions defined by an elementary cone containing the given direction, by the product of the solid angle ($d\omega$) of the cone and the area (dA) of the orthogonal projection of the element of the surface on a plane perpendicular to the given direction.

Radiance is the radiometric equivalent of luminance and is preferred for unambiguous measurement.

Note: In the defining equation, θ is the angle between the normal to the element of the source and the direction of observation.

Subjective brightness   The subjective attribute of any light sensation giving rise to the percept of luminous intensity, including the whole scale of qualities of being bright, light, brilliant, dim, or dark.

Note: The term *brightness* is often used when referring to the measurable "photometric brightness." While the context usually makes it clear as to which meaning is intended, the preferable term for the photometric quantity is luminance, thus reserving brightness for the subjective sensation.

Time of persistence   The time which elapses between the instant of removal of the excitation, and the instant at which the luminance or radiance has dropped to a stated fraction of its initial value.

APPENDIX 3.2
Conversion of Candelas to Footlamberts

Table A3.1 is an example of the conversion of candelas (or microcandelas) to footlamberts. It also shows the impact of segment area (last column) on display surface brightness.

Table A3.1.   Typical Performance Characteristics of LED Numeric Displays

	Monolithic numeric red	Reflector numeric		
		Red	Orange	
Character height, in	0.1	0.3	0.3	0.56
Segment area	1.6×10^{-3} cm²	27×10^{-3} cm²	27×10^{-3} cm²	93×10^{-3} cm²
Luminous intensity (microcandela at 10 mA dc)	275	250	750	700
Surface brightness (fL at 10 mA dc)	500	24	75	23

SOURCE: M. G. Craford, *Proc. of the Soc. for Inf. Disp.*, **18** (1977), p. 159.

4

Gas Discharge Displays

4.1 INTRODUCTION

The gas discharge displays area is a mature technology, and the displays have bright, pleasing digits. They have the disadvantage of being expensive to drive with semiconductor processors because of their high voltage requirements.[1,2] The display is based on the familiar property that neon gas radiates a number of yellow and red spectral lines when it is electrically excited. The excitation method may be dc or ac and leads to at least two separate device categories: dc discharge and ac plasma displays. The ac display is not widely used commercially and will be discussed briefly; the chief interest will be the dc display.

Discharge fundamentals will be considered and this discussion will lead to device parameter, driving method, and design factor considerations. Then specific displays and design examples will be considered.

4.2 GAS DISCHARGE FUNDAMENTALS

Gas Discharge Geometry

To understand the gas discharge display, the elementary gas discharge is considered.[3,4,5,6] If a variable power supply is connected to two electrodes of a gas-filled tube through a resistor (Fig. 4.1), it will be possible to strike a glow discharge between the anode and cathode. The regions of the glow are shown in Fig. 4.2. If the anode is moved toward the cathode, the glow region around the cathode will not change, while the positive column will shorten correspondingly until the anode reaches the Faraday dark space, at which point the positive column will disappear. The cathode glow is essential for the discharge and is the source of the light for

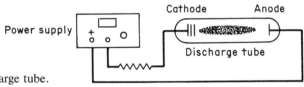

Fig. 4.1. Elementary glow discharge tube.

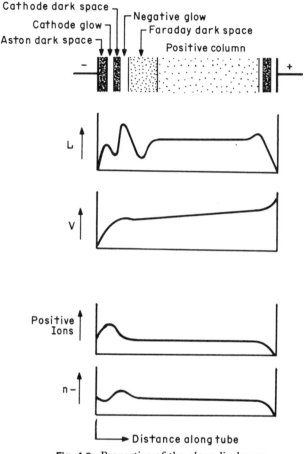

Fig. 4.2. Properties of the glow discharge.

many gas discharge displays. (Applications of the positive column include its use as a source of uv light in fluorescent tubes, gas lasers, neon signs, and some specialized displays.) The cathode glow of a neon-3% argon mixture with mercury added to retard sputtering has the familiar neon orange color with a bluish tinge around the edges. The positive column is a bluish color. The discharge is sustained by electrons from the cathode, which is cold (it rarely will exceed 40°C under the severest operating conditions in a room ambient). The sustaining electrons are secondary emission from the cathode and are primarily the result of positive-ion bombardment and secondarily of uv or radioactive bombardment. As such electrons are accelerated through the high field region of the cathode, they ionize the gas and produce the light. Once the discharge is struck, the segment voltage may be fixed and the current varied by means of the load resistor R_L. If R_L is decreased, the segment current is increased and the glow will spread over the cathode electrode. The current density remains constant in the normal glow region. This region of operation is known as that of the normal glow. Once the cathode is covered by the glow, further current increase raises the current density; this region is the abnormal glow region.

Current-Voltage Relationships of the Discharge

As the voltage of the circuit of Fig. 4.1 is increased, no current will flow until the breakdown, or sparking voltage (V_K) is reached. At this point a current will flow limited by the load resistance and the tube dynamic impedance. This behavior is shown in Fig. 4.3. A large-load resistor will lead to a small tube current; the glow will not cover the cathode in this the region of normal glow. The glow size adjusts to maintain constant current density (Fig. 4.3c). As the current is raised, the glow will

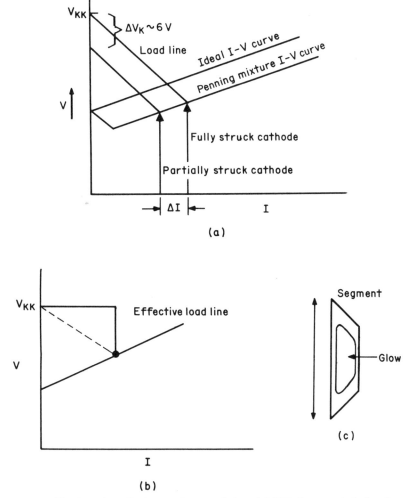

Fig. 4.3. Further glow discharge characteristics. (*a*) I-V characteristic for the circuit of Fig. 4.1. The incomplete struck cathode effect (*c*) is accentuated by metastables and electron attachment. (*b*) constant-current operation; (*c*) a-segment with struck glow showing incomplete segment coverage. There will be a minimum ΔV above the minimum V_{KK} required to completely fill a well burned-in cathode.

spread over the cathode until it is completely covered. This is the minimum current for a display tube. When a vacuum tube or transistor curve tracer is used to determine the i-v curve, the applied voltage has the shape of the solid curve of Fig. 4.4*a* without the tube and the shape of the dotted curve with the tube in place. The appearance of the trace is shown in Fig. 4.4*b*. Likewise, the behavior for a pulse source is shown in Fig. 4.5. The behavior of the discharge may be approximated by the equivalent circuit of Fig. 4.6, and the elements are explained there. The terminology is summarized in Table 4.1. One application of the equivalent circuit is to approximate the tube waveforms, obtain timing relations, and determine approximate operating points, such as described in Appendix 4.1.

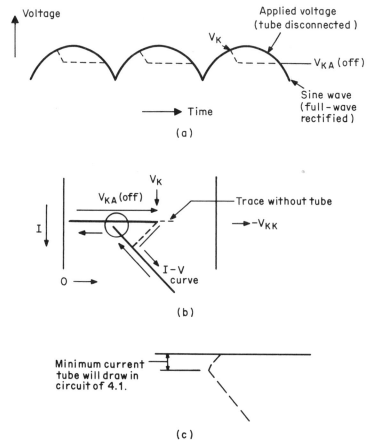

Fig. 4.4. Interpretation of curve tracer tests of a gas discharge display. (*a*) Curve tracer waveform (solid) and tube voltage (dotted) with the load resistor set internally to the curve tracer. Voltage is for oscilloscope attached across tube. (*b*) Curve tracer presentation showing parameter representation and direction of trace. (*c*) Enlarged area from (*b*).

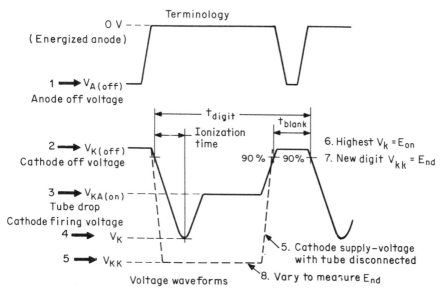

Fig. 4.5. Display applied-voltage waveforms showing terminology. *(After Ref. 12.)*

Applications to the Display

The glow discharge is used in a display as shown in Fig. 4.7. The cathodes are shaped as segments of a 7-segment display. The anode is a metal screen or transparent conductor spaced about a millimeter away. The interelectrode space is filled with a neon-argon gas mixture and the leads are brought outside the package envelope. The figure shows the *f*-segment switched on; the other cathodes are off.

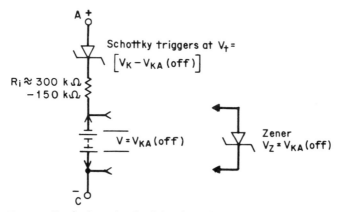

Fig. 4.6. Equivalent circuit of the glow discharge. The battery (or zener diode) represents the voltage V_{KA}(off) at which the discharge extinguishes. R_i is the plasma resistance and is set by the gas pressure.

Table 4.1. Some Gas Discharge Terminology*

V_{KK}, **Cathode Supply Voltage**—The total voltage with respect to an energized anode terminal, applied to the cathode terminal through the ballast resistor.

I_K, **Cathode Current**—The current into the cathode terminal. Negative values indicate current out of the terminal.

$V_A(\text{off})$, **Anode Off-state Voltage**—The voltage at the terminal of an unselected cathode with respect to an energized anode terminal.

$V_K(\text{off})$, **Cathode Off-state Voltage**—The voltage at the terminal of an unselected cathode with respect to an energized anode terminal.

t_{digit}, **Digit Time Period**—The time interval starting when one digit or character is selected and ending when the next digit is selected.

t_{blank}, **Segment Blanking Interval**—The time interval between successive address pulses provided to ensure glow extinction.

$V_{KA}(\text{on})$, **Segment On-state Voltage**—The voltage at the cathode terminal with respect to the anode terminal after the segment is fired.

Ionization Time—The time interval between the initiation of conditions for and the establishment of conduction.†

Initial Ionization Time—The ionization time following nonoperating storage.

New-Digit Ionization Time—The ionization time for a segment in a digit adjacent to a digit that was operating during the previous cycle.

Reionization Time—The ionization time for a segment in a digit adjacent to a digit that was operating during the previous cycle.

*These symbols and terms were selected in accordance with the rules of the International Electrotechnical Commission (Publication 148) and the Electronic Industries Association JEDEC Publication 77).

†See IEEE Standard 100, p. 297.

SOURCE: Ref. 14.

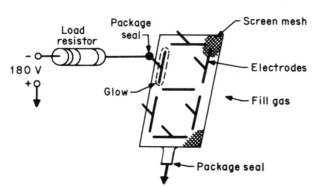

Fig. 4.7. Application of glow discharge to display tube.

**Table 4.2. Metastable, Radiation, and First
Ionization Potentials for the Noble Gases***
(All energies are in electron volts.)

Atom	V_m	V_i	V_r
He	19.8	24.58	21.21
Ne	16.62	21.58	16.85
Ar	11.55	15.77	11.61
Kr	9.91	14.01	10.02
Xe	8.32	12.14	8.45

*Note, for example, that the Ne metastable has a higher
energy than the ionization energy of one argon electron
(boxes).

SOURCE: Ref. 17. Copyright 1965 by Bell Telephone Laboratories. Reprinted by permission.

Fill Gases

Displays are normally filled with a Penning mixture. The Penning mixture is of a
suitable light-emitting gas such as neon and a gas with a lower ionization potential
than its metastable (or long-lived) state, such as argon (Table 4.2). A small amount
of a radioactive gas such as the krypton 85 isotope or tritium may be included to
provide initial ions to aid starting.

A small amount of liquid mercury is usually added and its vapor, in the gas and as
an electrode coating, reduces cathode sputtering and extends tube life (Fig. 4.8).

Once the gas filling and cathode material are established, the Paschen law states
that, to the first approximation, the breakdown voltage and running voltage are

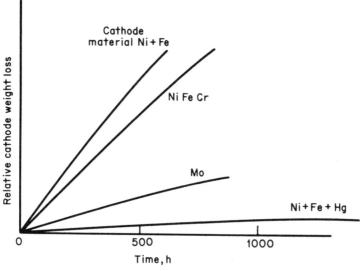

Fig. 4.8. Relative cathode sputtering in Ne:Ar (99:1) discharge showing
the striking reduction with mercury (Hg) addition. *(Adapted from Ref.
16.)*

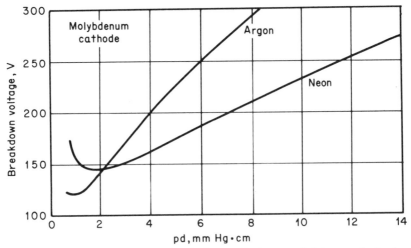

Fig. 4.9. The breakdown voltage for devices with planar molybdenum electrodes and neon and argon fillings. *(From Ref 17. Copyright 1965 by Bell Telephone Laboratories. Reprinted by permission.)*

functions of pd only (Figs. 4.9 and 4.10). A typical display tube would have a spacing d of 1 to 1½ mm and a gas pressure p of 60 to 100 mmHg (torr). The pd values would range from 6 to 15. Choice of pressure is made on two bases: (1) low pressures (below the pd minimum) allow the discharge to become fuzzy and close the gaps between adjacent electrodes to provide a continuous character; (2) pres-

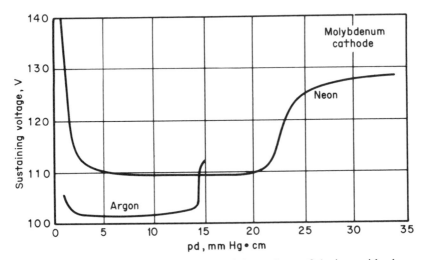

Fig. 4.10. The normal-glow-discharge sustaining voltage of devices with planar molybdenum electrodes and neon and argon fillings as a function of pd. In obtaining the curve for neon p was held at 50 mmHg, and d was varied. In obtaining the curve for argon p was held at 25 mmHg, and d was varied. *(Adapted from Ref. 17, p. 551. Copyright 1965 by Bell Telephone Laboratories. Reprinted by permission.)*

sures above the pd minimum reduce sputtering, which can depend on p^3, and also reduce the effects of gas depletion through ion pumping effects. As pd is decreased below 5, the discharge is constrained by the spacing d and the current density changes, whereas for pd greater than 5, it is a constant. In practice, values greater than 5 reduce the effect of manufacturing variations in spacing tube-to-tube or within a tube on the current density or the brightness.

In practice, note also that if the tube has different-size cathodes, the current for each must be varied proportionally with the area to maintain the current density and hence the brightness per unit area. Sometimes parameters are known for one size of tube and it is necessary to predict their behavior for a different-size tube that is constructed the same way. Use of the scaling relations of Fig. 4.11 will sometimes allow estimates of operating parameters for one tube to be based on those of another.

Penning mixtures are the heart of low-voltage displays; when a few tenths percent of argon is mixed in neon, the starting voltage (V_K) and the running [sustaining or V_{KA}(on)] voltage are sharply reduced (Fig. 4.12). The results are shown for molybdenum cathodes. Other relations and cathode material values are given in Appendix 4.2.

Light Emission

The spectrum of a dc flat-panel tube from a TI 3500 calculator is shown in Fig. 4.13. This tube is filled with a neon-0.3% argon mixture. (Note that the spectral intensities in the figure must be corrected for the photomultiplier response.) All the light is sharp line spectra from neon (Fig. 4.14). Lines from argon are not seen because of its low concentration. As seen in the figure, the spectral lines arise from transitions from the ten $3p$ states to the four $3s$ states (also known as the $2p$ and $1s$ states in

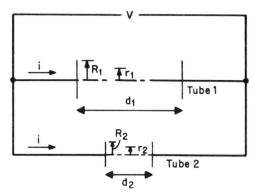

Fig. 4.11. Scaling relations between two tubes where the voltage and current are held constant. J is the current density, T is the temperature, and p is the pressure. *(From Ref. 18.)*

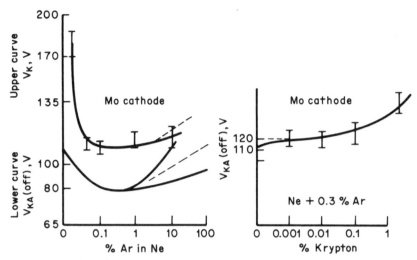

Fig. 4.12. Variation of tube parameters with percentage argon in neon on the right. The left curve shows the insensitivity of V_{KA}(off) to small amounts of radioactive krypton ([85]Kr) which is often added to aid starting. *(From Refs. 4 and 19.)*

Paschen notation).[7] Transitions from the $4s$ states are in the infrared and are not seen. Apparently the lack of lines higher than 20 eV (588 nm) in energy is because electrons scatter frequently in the cathode fall region and most never gain more than 20 eV of energy. The population of the $3p$ levels will be a function of the location in the gas, its pressure, and its temperature.

The color of the discharge will vary from region to region. The cathode glow of neon is yellow, while the negative glow (the brightest part) is orange because the electron streams in these two regions have different properties. The higher-energy electrons are found nearest the cathode and thereby excite the optical electron of neon more strongly. The discharge color nearest the cathode will then be more nearly yellow than the portions farther away. Where mercury is added to retard sputtering, a bluish tinge at the edges may be seen. Other gases will have different colors (Table 4.3).

Alternately, colors other than red or orange can be obtained by means of phosphor-coated cells activated by means of uv light from a helium or xenon discharge, or by particle bombardment from the discharge. The former procedure is expected to have a longer phosphor life since it is somewhat similar to fluorescent lamp operation.

AC Plasma Discharge[8]

The ac plasma is used in flat-panel dot-matrix displays because it has memory, which leads to improved brightness at high cell density. It also has the advantage of not requiring load resistors. It has the disadvantage of needing a driver for each row and column and requiring relatively high power-supply voltages. It is constructed with an insulating layer over the cathode and anode which are charged each half

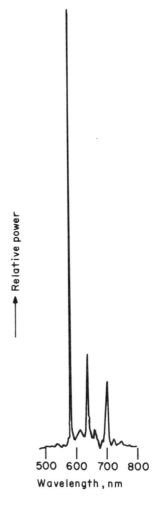

Fig. 4.13. Radiation spectra from a Ne:Ar (99.5:0.5) discharge. The 520 photomultiplier correction factors for the major peaks are (increasing λ) about 1:1½:3.

cycle of the driving waveform. The fill gas will usually be a Penning mixture with a small amount of nitrogen added to aid wall charging. The driving waveform is alternated about 130 V around a ground reference and is set equal to about V_{KA}(off) for the gas-insulator combination. (See Table 4.1 for the terminology.) A plasma cell is activated by adding a turn-on pulse after the sustain waveform (Fig. 4.15). The width determines the discharge intensity as based on the cell's previous history and that of its neighbor. A useful design tool is the voltage transfer curve, which relates the change in wall voltage to the wall voltage before firing. The wall voltage is explained in Fig. 4.15b. It is seen to be the driving voltage $V_{KK} + V_{CAP}$ where V_{CAP} is the voltage across both insulating films and is equal to $V_K - V_{KA}$(off) for a sustaining voltage of V_{KA}(off). Figure 4.16 shows a representative voltage transfer curve. Note that over the steady sustain region the change in wall voltage is equal to the initial gas voltage. Prior to this region V_{CAP} opposes V_{KK} (see Ref. 9 for more details where desired). A line drawn with a slope of 2 (for a $+V_{KK}$ to $-V_{KK}$ change)

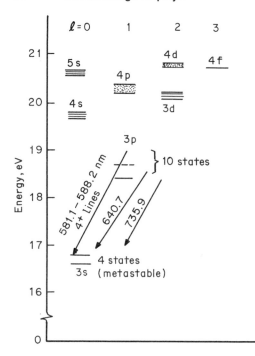

Fig. 4.14. The energy levels of the lower excited states in neon showing the transitions leading to "neon orange."

will intersect this curve in three points, of which two are stable. The operating margin is the voltage difference between tangent extreme slope 2 lines (points *A* and *B* of the figure) and is typically 10 V for new tubes. Detailed application data must be supplied by the manufacturer, of course.

4.3 THE GAS DISCHARGE DISPLAY

Display preliminaries were discussed in Sec. 4.2, under Applications to the Display. Construction techniques are considered here since they determine to some extent

Table 4.3. Glow Discharge Colors

Gas	First cathode layer	Negative glow	Positive column
Air	Pink	Blue	
Hydrogen	Brownish-red	Pale blue	Pink
Nitrogen	Pink	Blue	Red
Oxygen	Red	Yellowish-white	Pale yellow (pink center)
Helium	Red	Green	Red to violet
Argon	Pink	Dark blue	Dark red
Neon	Yellow	Orange	Brick red
Krypton		Green	
Xenon		Olive green	
Mercury	Green	Green	Greenish

SOURCE: Ref. 13.

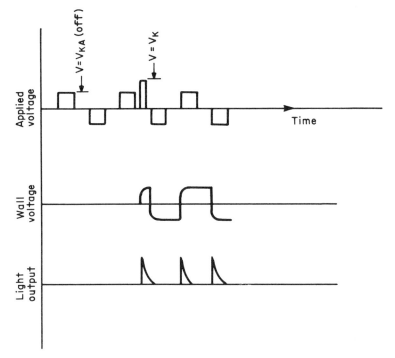

Fig. 4.15a. Ac plasma driving waveform, wall voltage, and light output. Turn-on occurs with a turn-on pulse and its width limits the plasma current.

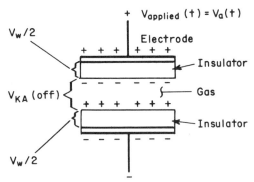

Fig. 4.15b. The wall voltage V_w. When there is no discharge and $V_a(t) = 0$, the wall charges to oppose current flow through the gas. The voltages then add to zero:

$$\frac{V_w}{2} + \frac{V_w}{2} + V_{KA}(\text{off}) = V_a(t) = 0$$

The wall voltage is $V_w = V_{KA}(\text{off})$. In the event a turn-on pulse is applied, the wall charges to a sustaining voltage.

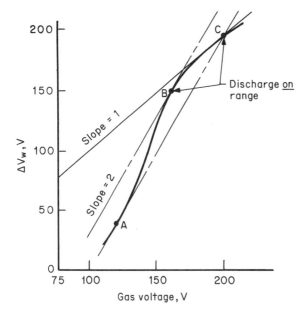

Fig. 4.16. The ac plasma voltage transfer curve. *(Adapted from Ref. 20.)* Change in wall voltage V_w related to the voltage across the gas.

the allowable display configurations. Applications and failure mechanisms are also considered.

Gas Discharge Display Applications

DC gas discharge displays are principally used for applications requiring high brightness (100 to 200 fL) and, in particular, large sizes [up to 2.50 cm (1 in)]. They are used for games, cash registers, point-of-sale (POS) devices, gas pumps, and automobile sockets. Dot-matrix applications involve up to 4 rows of 20-character 5 × 7 displays 0.80 cm (⅓ in) high. Typical luminances are 50 fL (without the filter, which may or may not be used).

The ac displays have more elements (up to 200,000 is typical) and brightness values in the neighborhood of 20 to 30 fL. The colors are neon orange but may be filtered between red and yellow by weighting the contributions of the groups of yellow and red spectral lines.

The xenon-argon–filled displays mentioned earlier can activate green and white phosphors by means of the xenon radiation, but life data is not readily available.

Construction Techniques

Construction techniques will be suited to the cathode type—either solid or film. The solid cathode may be nickel or molybdenum, but generally it is made of the cheaper stainless steel. The mercury which is added to retard sputtering will modify the electrical characteristics accordingly. Figure 4.17 shows a construction which may include a recess behind the cathode. Mercury may collect in this recess rather than on the cover glass, where it would be obstructive, or on the cathode, where it would cause bright spots. The anode is a metal mesh or transparent indium or tin oxide coating on the cover glass. Contact to the thin-film anode is by means of a spring

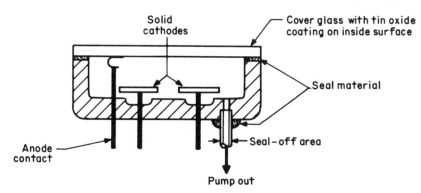

Fig. 4.17. Cross-sectional view of a popular construction technique for the gas discharge display which uses the solid cathode.

clip. There are generally two or three characters to an envelope, which must be interconnected externally for multiplexing by means of a printed circuit (pc) board. Cathode shapes are generally limited to a simple stampings or lead frame geometries.

The film cathode may either be a screened thick film or an evaporated thin film. Thick films are most common; the tooling is the least expensive and may be completely changed for about the price of a medium-priced automobile. Format variations will cost about half this amount. Again construction will incorporate mercury for sputtering retardation, but problems may occur with spots. A getter which is selective to the rare gas filling and to mercury is used in both tube types; it is primarily used to getter oxygen, nitrogen, CO_2, and water vapor as outgassing products.

Figure 4.18 shows details of construction of the thick-film cathode display; from 2 to 20 characters may be incorporated in a single envelope. Interestingly, many of the fabrication costs will not be a strong function of digit number since the major

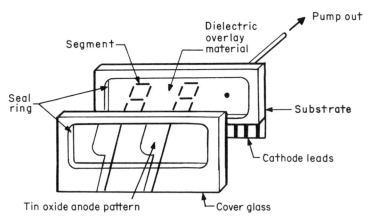

Fig. 4.18. Blowup of screened cathode construction gas discharge display.

material costs can be in the substrate, cover, and seal ring, and the major assembly costs are relatively size insensitive.

Hard contacts (as opposed to pluggable contacts) may be provided to the thick-film ceramic package by epoxying spring contacts in place.

Forming

Once the display is fabricated, the cathodes and the gas are cleaned by means of a forming procedure. A temporary mercury-coating procedure may also be done. The forming is carried out at a high current density and at a voltage high enough to ensure positive striking in a contaminated gas and with contaminated cathodes [for example, a small amount of oxygen will raise the striking voltage of Ne-Ar (99.5:0.5) to 300 V]. Slight cathode deposits, introduced during assembly, will sputter off upon proper choice of cycle and some gas cleanup may occur by means of sputter ion pumping. A typical cycle would be about 4 times current and a V_{KK} of about 350 V for a few minutes followed by 1 to 2 times current for a few hours. Then a period of operation under normal conditions—possibly with a 350 to 450°C bake cycle to distribute the mercury (not required in well-manufactured tubes)—is used to establish a uniform coat of mercury on the cathode. This coating process is not stable beyond about three months' storage and may be more conveniently included in the end item burn-in process. Note that this instability may be a detrimental factor in incoming tests on stale date code tubes or with the end item product which has been on the shelf for some time. For this reason, drive circuits are often rated to aid in-circuit cathode formation after periods of disuse. The display manufacturer will be able to furnish details concerning his particular tubes and the stability of his processes in this regard.

Failure Mechanisms and Temperature Considerations

Failure mechanisms for dc screened cathode gas discharge displays are given in Table 4.4. Lists of mechanisms for other types of displays are generally not

Table 4.4.   Some Flat Panel Failure Mechanisms

Symptom	Possible cause
1. Fading to severe fading (tiger striping)*	Selective burn-in of small cathode areas due to gas or cathode contamination
2. Sputtering:* Anode discoloration Fuzzy segments Severe mottling	Mercury exhaustion [or low temperature (<0°C) operation]
3. Slow-to-enter digits:† Flickering digits	Getter exhaustion (gas contamination)
4. Flickering display, segments, or digits‡	Contact failure due to temperature cycling or on-off cycling
5. Mercury spots§	Small drops of mercury on cathode

   *Items 1 and 2 will exceed 9,000 to 12,000 hours in well-manufactured tubes.

   †Item 3 can be in the 1000 to 1500 hour range in commercial tubes for typical specifications and thermal cycles and double-type drive circuit 2000 to 3000 hours for constant-current driver.

   ‡Item 4 varies from 3 to 5% failure per 1000 hours to magnitudes less depending on contact system.

   §Item 5 may not be objectionable.

Table 4.5.   Some End-of-Life Criterion

Criterion	Typical times
Meets new display acceptance test criterion.	10,000 hours or greater in well-manufactured tubes.
Readable display.	Greater than 30,000 hours in well-manufactured tubes.

available. Tube life for given failure modes based on past experience and the hypothesized source of the failure are also listed. The actual times for any particular tube will, of course, depend on the specific tube process, drive circuit, end-of-life or failure criterion, operating temperatures, drive current, and other factors. Two end-of-life criteria are given in Table 4.5 and show how dependent tube life is on such specifications. (Lifetime is of course a mean time; gas displays exhibit the typical bathtub failure curve: 3 or 4% early failures, very low constant failure rate for the lifetimes mentioned above, and then rapid loss of population.)

Table 4.6 shows how cosmetic appearance might be specified; tubes that don't meet this criterion would be said to have failed by the first end-of-life criterion of Table 4.5. Expected life data for a solid cathode tube manufacturer is shown in Fig. 4.19 as a function of duty cycle at ambient temperature.

Electrical parameters also change with time; a particularly critical one for screened cathode tubes is the overvoltage ΔV required to enter a new digit in less than $\frac{1}{3}$ second (where $E_{ND} = V_K + \Delta V$). The variation of E_{ND} with time for a tube with no getter (1974 PP* II) and one with a special dielectric getter is shown in Fig. 4.20.

Environmental stress can be a factor in increased failure rate. However, little is known about the effects of various environments on life. It is known that item 2 of Table 4.3 is accelerated by temperatures below 10°C and item 3 is accelerated by temperatures above 55°C. Details are generally not available from manufacturers. Testing programs will be required where high temperatures are routinely encountered, such as for oven controller displays. Discharge-tube temperature will be slightly higher than ambient temperature when in operation but usually not much. The 10-digit display in the now-discontinued TI 3500 calculator operates about 58°C in a 55°C ambient.

*Trademark of the Burroughs Corp.

Table 4.6.   Cosmetic Specification

	Specification
Spatial uniformity:	
Bright spots	No more than twice as bright
Broken segments, faded segments, partial segments	Differential brightness less than twice. Not more than 3 minor or no major per display
Missing segments	None allowed
Temporal uniformity:	
Flickering segment	Nondistracting
Slow-to-enter	$<\frac{1}{3}$ s max

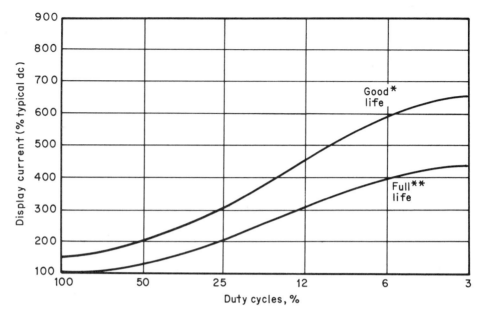

Fig. 4.19. Rerated life expectancy of a solid cathode gas display as a function of current and duty cycle. *(Adapted from Ref. 21.)*

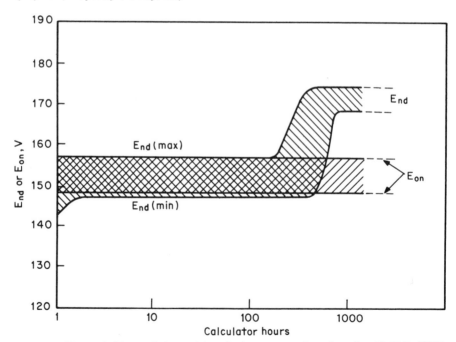

Fig. 4.20. E_{nd} and E_{on} variation with calculator operating time for 10-digit (1974 vintage) Panaplex units. (Panaplex is a registered trademark of Burroughs, Corp.) *(After Ref. 12.)*

Electrical parameters are insensitive to temperature in the range 0 to 100°C[12] and generally can be ignored in practical circuit designs.

Special Provisions and Requirements

Tubes in the dark that do not have provision for initial starting may require many minutes before arrival of a cosmic ray to provide starting. Reliable starting is generally provided by addition of a small amount of radioactive material.

Scanned tubes require reliable reset to start the first scan character, and often a separate keep-alive cell is provided for this purpose. Keep-alive cells will also reduce the overvoltage required for reliable firing and will minimize long-term contamination effects. The effect of a keep-alive cell near an indicator cell (ic) on the ic's i-v curve is shown schematically in Fig. 4.21. The manufacturer's data sheet will specify keep-alive operating requirements.

Tubes with keep-alive circuits may require tighter V_{KK} tolerances to prevent undesirable arcing; testing at line-voltage or battery-supply extremes should be planned to establish these requirements.

Tubes without keep-alive circuits, on the other hand, may have a narrow multiplexing rate range and also restricted between-character blanking times required to prevent slow turn-on, incomplete turnoff, or arcing problems.

Generally speaking, the use of well-regulated power supplies and low-impedance drive circuits will minimize the effects of tube-to-tube variations once operating ranges have been established.

Radio-frequency interference (rfi) may be a problem with the multiplexer drive circuits or power supplies and the usual precautions are necessary. A small AM-FM radio receiver may prove useful for prototype sweeping. The gas tube may also be a

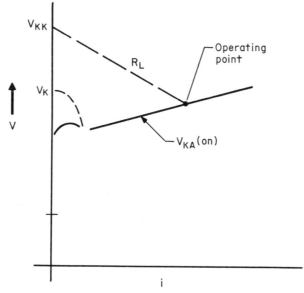

Fig. 4.21. Discharge i-v curve showing V_K reduction by keep alive or residual ions (lower dark curve).

source of rfi, but such cases are rare. Shielding may be provided by means of a transparent conductive coating over the tube face, if it should be required.

4.4 DC DRIVE CIRCUITS

Anode Drivers

Anode drivers in multiple-digit multiplexed tubes must sink currents that vary from 1-segment current to n-segment currents, depending on the number of segments in the tube. Ideally, they will be able to return the anode to ground before the cathode is switched and then return it to $V_A(\text{off})$ before the next anode is energized.

A typical anode driver is described in Table 4.6.

Segment Driver Circuits

A variety of segment drive circuits are in use. A few are described in what follows. Integrated-circuit drivers are also available (Table 4.7).

Segment drive waveforms are shown in Fig. 4.22 for some different conditions to aid in interpreting multiplexed display waveforms. Figure 4.22a shows the segment waveforms of a calculator with an initial zero. At the left the calculator is turned on. Pulses are applied to the a-segment of digit 1, until a second or so passes and the segment fires as a result of a-segment secondary emission by beta particles from krypton 85 decay.

Table 4.7. DC Gas Discharge Drivers

Descrip-tion	Input compatibility	Power supplies	Drivers per package	Device type	Additional features
High-voltage BCD-to-7-segment decoder /cathode drivers	TTL	5 V	7	SN75480	Outputs regulated to ensure constant brightness Blanking and ripple-blanking provisions High off-state breakdown voltage (120 V typical) Designed for 7-segment displays such as Beckman and Panaplex II*

	TTL, MOS, CMOS	Variable from 4.75 to 15 V	7½	SN75484	Same features as the SN75480 plus:
					  Decimal point provided
					  Latches to hold BCD information
					  Lower supply-power requirements
					  Higher output-voltage breakdown capability

Anode driver	MOS	$V_{EE} = -55$ V, $V_{BB} = -18$ V	6	SN75481	13-mA output capability Designed for time-multiplexed displays such as Panaplex II*

*Panaplex II is a trademark of the Burroughs Corporation.
<small>SOURCE:</small> Ref. 15.

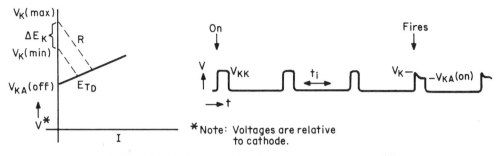

*Note: Voltages are relative to cathode.

(a) Initial ionization time, t_i — time to enter first ⌐

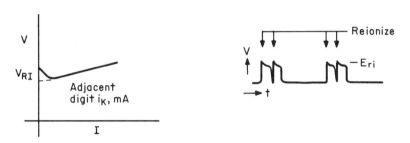

(b) Reionization time – digit period leading to flicker

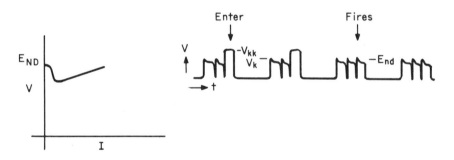

(c) Entry ionization time — time required to enter new digit

Fig. 4.22. Firing voltages and firing times for cold cathode tubes.

In Fig. 4.22b the a-segment is on in digits 1 and 2. In Fig. 4.22c, the a-segment in digit 3 is energized and fires on the third MUX cycle.

Transistor Drive. One form of transistor drive is shown in Fig. 4.23. The analysis and waveforms are shown there. Note that the transistor has a large collector-base breakdown voltage requirement.[22]

Doubler Circuit. One approach to reducing the drive-transistor voltage is the voltage doubler circuit of Fig. 4.24. Here the transistor only has to swing about half the display V_{KK} requirement. Although this circuit is widely used, it is not as tolerant of tube-to-tube variations or variations over the life of a tube. Some advantages of this circuit are developed in Appendix 4.1.

Special circuits have been described[10] that allow further reduction in voltage

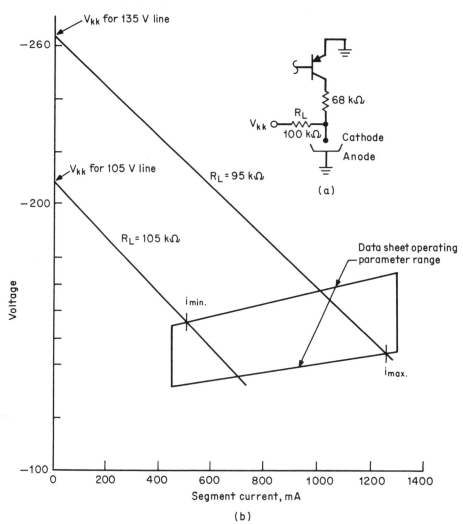

Fig. 4.23. Transistor drive circuit for gas discharge display. (*a*) Circuit; (*b*) load line analysis of circuit a. (*R. Whitaker, Texas Instruments Inc.*).

swing requirements and hence lower breakdown voltage drivers, but they are not popular because of the critical circuit tolerances.

Constant-Current Drive. Many of the disadvantages experienced with the doubler circuit can be eliminated by use of the constant-current driver, which is also analyzed in Appendix 4.1. Commercial driver chips are available to perform this function (Table 4.7). The use of the control diode described in the next section in conjunction with this circuit can reduce drive-transistor breakdown voltage requirements.

Control Diode. It is possible to buy two electrode tubes from neon bulb manufacturers which have the same i-v characteristics as the display tube. These are

Anode (digit) select drivers

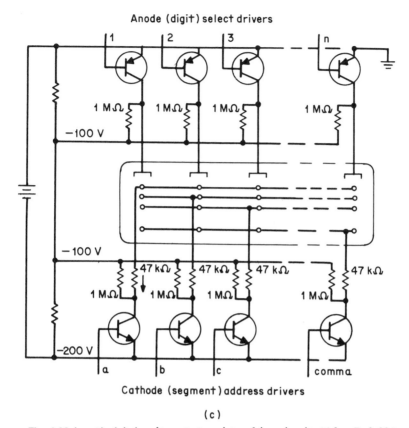

Cathode (segment) address drivers

(c)

Fig. 4.23 (cont.). (*c*) An alternate transistor drive circuit. (*After Ref. 23.*)

available in quantity for a few cents a piece and allow an *n*-segment, multiplexed display to be driven with *n* bulbs, as shown in Fig. 4.25. Such an arrangement has also been described for diodes contained within the display envelope.[12]

The circuit operation may be explained as follows by considering the *a*-segment; in digit 1 it is on and in digit 2 it is off.

When digit 1 is MUXed, switch *a* is open and segment *a* fires. During the anode (digit) blanking time, the segment *a* switch is closed and the control diode fires. Its $V_{KA}(\text{on})$ is less than V_K for segment (by choice of gas pressure or current density as required), and segment *a* remains unstruck when anode 2 is selected. Note that the MOS chip segment outputs must be inverted to obtain proper control. V_{KK} must be well-regulated and at only about V_K to minimize $V_a(\text{off})$ to 20% of V_K.[1,2] A design example is given in Fig. 4.26.

In the event proper blanking is not available, the control diode anodes may be pulled positive (or zener diodes used in the digit anodes) to obtain control.

Multiphase Clock Circuits. Dot-matrix and multiple-bar displays (for analog applications) often use a multiphase clock to transfer the glow from one bar to another in a manner analogous to that used in a charge-coupled device (CCD). The

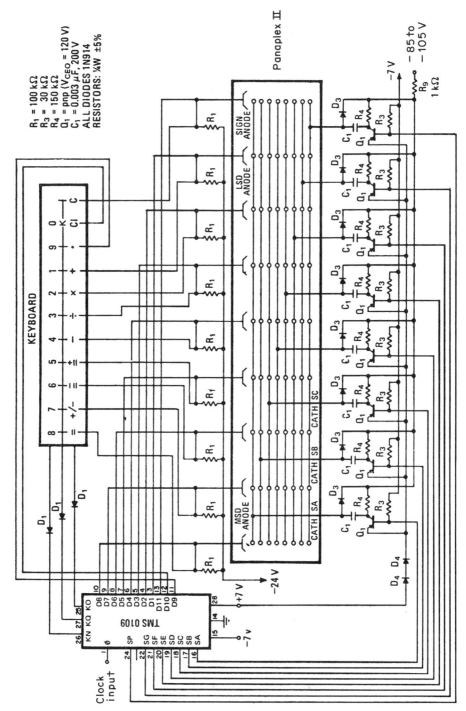

Fig. 4.24. Calculator interface using doubler circuit. *(From Electronics, March 1973, by permission.)*

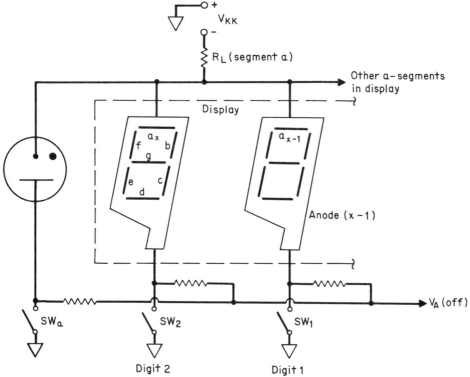

Fig. 4.25. Low-voltage control of a gas discharge display by means of a control diode (CD). There is one control diode for each set of common segments. The a-segment control diode is shown. The switching transistor, represented by swa, must switch only V_A(off), which can be less than 25 V. SW1 and SW2 are alternately selected by the multiplexer. If SWa to SWg are open, all 8's will be displayed. If swa is closed slightly before swl, for example, the a-segment control diode's charging waveform will lead to that of a_1-segment. The control diode will fire and its voltage drop, V_{KA}(off), will now be less than the a-segment firing voltage, V_K, and the a-segment is held unconducting.

An alternative approach where an unbalanced CD pulse (segment select) is not available from the MOS chip is to use a positive reference voltage for the CD switch as supplied by a zener or other reference supply.

principle is shown in Fig. 4.27. As a rule, these devices are sold with random-access memories (RAMs), power supplies, and drivers as a unit. Application data are available from the manufacturer when they are not.

4.5 AC DRIVE CIRCUITS

In general, circuits for ac panels are more complicated than those for dc panels. Again the reader is referred to the vendor for applications data. An introduction to such circuits may be found in Ref. 11. Integrated-circuit drivers are summarized in Table 4.8.

Purpose: Design Procedure

Designer chooses parameters:
1. Ik for brightness-chip diss. tradeoff
2. t digit for flicker V_{kk} tolerance tradeoff
3. V_{kk} and R_L then fixed
4. V_{kk} and or R_L worst and best cased
5. Ik worst and best case est. (in spec√)
6. V_{ak} (off) to prevent extraneous glow. chosen from worst case V_{kk}, V_{ka} (on), (ik)
7. t blank chosen to prevent arcing at max V_{kk}

Control diode tube
8. MOS chip chosen with small or no cathode blanking

(a)

Example: 8 or 10 digit 0.4 inch character

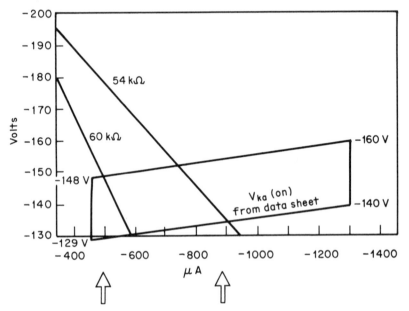

- Nominal 500 μ A chosen from past experience with brightness,
- t Digit = 250 μ s chosen because 13-25 μ s t blank and no flicker, (Many MOS chips limited to 300 μ s MAX)
- Min. V_{kk} chosen to be 180 V new digit entry safety factor Max. V_{kk} 195 V arcing safety factor R_1 chosen from load line analysis V_a (off) at -20 V OK? Rule check (V_{KA} (on) + 125 V) = V_a (off) > -145 + 125 = -20 V no production margin use -25 V
- V_{kk} < -210 and t blank < 10 μ s arc rule check
- Use -25 V on control diode anode since C.D. characteristics are the same as segment. (Note: C. D. may have 10%-15% greater area than segment to ensure positive control and improved life)

(b)

Fig. 4.26. (*a*) Gas discharge display. (*b*) Example of (*a*).

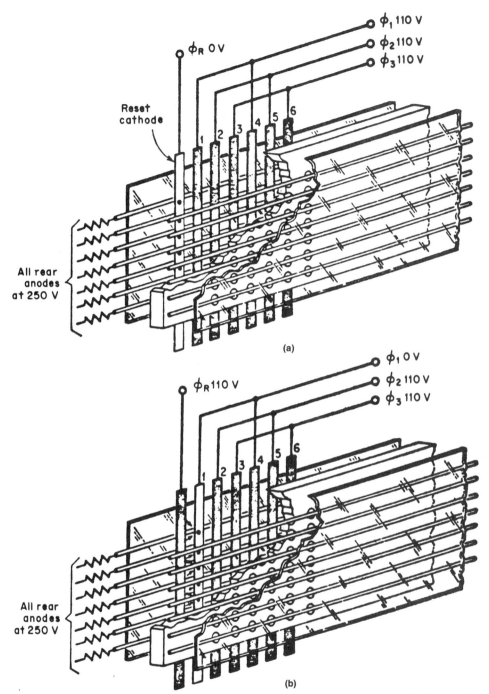

Fig. 4.27. (*a*) The use of priming to effect scanning with only three drivers in the Burroughs "Self-Scan" display. First step: A discharge is established between the reset cathode and the rear anodes. (*b*) The cathodes of phase 1 are dropped to 0 V, while the reset cathode is raised to 110 V. The discharge transfers to the first cathode. The other cathodes of ϕ_1 are too far away to be primed by the discharge at the reset cathode. (*c*) ϕ_2 is dropped to 0 V, while ϕ_1 is

(continued on page 86)

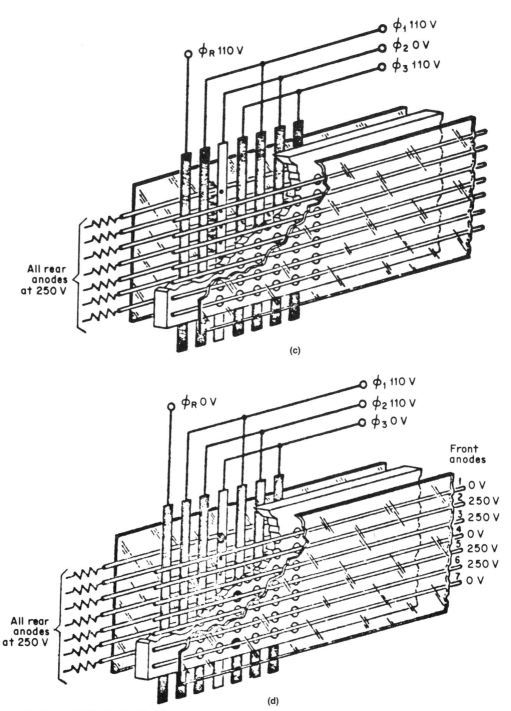

raised to 110 V. Again, the discharge transfers to the nearest low cathode. (d) By the same process, the discharge is transferred to the nearest cathode of ϕ_3. A glow appears in front only in those cells in which the front anode is high and there is an active discharge in the rear, that is, only in those front cells which are primed. *(From Ref. 22, W. J. Harmon, Electronics, 43, p. 120, Mar. 2, 1970. Copyright by McGraw-Hill.)*

Table 4.8. AC Plasma Drivers

Description	Input compatibility	Power supplies	Drivers per package	Device type	Additional features
Axis drivers	CMOS	$V_{CC1} = 12$ V	4	SN75426	Independent addressing of each gate for serial and parallel applications High input impedance (typically 1 MΩ)
		V_{CC2} variable from 40 to 90 V		SN75427	30-mA clamp diodes on output Switches 70 V in 1.2 μs AND driver (SN75426); NAND driver (SN75427)
	CMOS	$V_{CC1} = 12$ V	32	SN75500	High-speed serially shifted data input (4 MHz max) Fast output transitions (less than 150 ns) 25-mA output current capability Output short-circuit protection
		V_{CC2} variable from 40 to 100 V		SN75501	Static shift registers can retain data on all outputs of SN5501 indefinitely X-axis driver: SN75500 Y-axis driver: SN75501 (performs Y-axis sustaining function)

SOURCE: Ref. 15.

REFERENCES

1. An exception is the control diode circuit.
2. See Sec. 4.2, under Fill Gases.
3. A. Sobel, *Proc. of the Soc. for Inf. Disp.*, **18** (1977), p. 51.
4. J. R. Acton and J. D. Swift, *Cold Cathode Discharge Tubes,* Academic Press, New York (1963).
5. G. F. Weston, *Cold Cathode Glow Discharge Tubes,* Iliffe Books, London (1968).
6. J. D. Cobine, *Gaseous Conductors,* Dover, New York (1941).
7. Racah notation is given in I. I. Sobel'man, *Introduction to the Theory of Atomic Spectra,* Pergamon Press, New York (1972). Spectral details can be ferreted out of G. H. Dieke, "Atomic and Molecular Physics," *American Institute of Physics Handbook,* McGraw-Hill, New York (1957), chap. 7, and E. U. Condon and G. H. Shortley, *The Theory of Atomic Spectra,* Cambridge University Press, Cambridge (1963).
8. See Ref. 3 and other relevant papers in that same issue. Also see *IBM J. Res. Dev.*, **22** 6 (1978).
9. See O. Sahni, *Proc. of the Soc. for Inf. Disp.*, **18** (1977), p. 69, for more details.
10. J. Y. Lee and E. Lord, *Electronics,* March 1, 1973.
11. A. C. Cribbs and W. D. Petty, 1973 *IEEE Intercon Technical Papers,* Session 33, paper 33/4 (1973); E. S. Schlig and G. R. Stilwell, *Proc. of the Soc. for Inf. Disp.*, **18** (1977), p. 86; and T. N. Criscimagna, *Proc. of the Soc. for Inf. Disp.*, **17** (1976), p. 124.
12. E. G. Bylander, *Conference Record of 1976 Biennial Display Conference,* IEEE, New York (1976), p. 26.
13. G. Francis, "The Glow Discharge at Low Pressures," in S. Flügge (ed.), *Handbook der Physik,* Springer-Verlag, Berlin (1956), p. 58.

14. E. G. Bylander, "Components for Electrooptics," in C. A. Harper (ed.), *Handbook of Components for Electronics,* McGraw-Hill, New York (1977), chap. 5.

15. *The Display Driver Data Book for Design Engineers,* Texas Instruments, Inc., Dallas, (1977).

16. G. F. Weston, *Cold Cathode Glow Discharge Tubes,* Iliffe, London (1968), p. 335.

17. J. W. Gewartowski and H. A. Watson, *Principles of Electron Tubes,* Van Nostrand, New York (1965), p. 539.

18. G. Francis, "The Glow Discharge at Low Pressure," *Handbuch der Physik, Band XXII, Gasentladangen II,* Springer-Verlag, Berlin (1956), p. 70.

19. F. A. Benson, *Proc. IEE,* **106c** (1958), p. 82, and **108b** (1961), p. 82.

20. E. S. Schlig and G. R. Stilwell, *Proc. of the Soc. for Inf. Disp.,* **18** (1977), p. 86.

21. Beckman Instruments, Inc.

22. Private communication from R. Whitaker, Texas Instruments, Inc.

23. Cherry Electrical Products Corp.

24. W. J. Harmon, *Electronics,* March 3, 1970, p. 120.

APPENDIX 4.1
Analysis of Multiplexed Multiple-Segment Tube with Segment-to-Segment Leakage and Voltage Doubler or Constant-Current Drive

Although the voltage doubler circuit of Fig. 4.24 is widely used, it has important drawbacks. A 7-segment multidigit tube with such a segment driver will have slow turn-on of newly entered digits if segment-to-segment shunts of 3 MΩ or less exist. The slow turn-on, which is an increase in the number of applied pulses before firing, is a result of reduced-applied-voltage rise time and also reduced applied voltage (V_{KK}). This action may be understood by analysis based on the equivalent circuit of Fig. 4.6.

Figure A4.1 shows the arrangement of segments a and b in a 6-digit tube. Two of these digits are shown with digit-select switches and pull-down resistors in Fig. A4.1b. The equivalent circuit for two a-segments of Fig. A4.1 is shown in Fig. A4.2 where the cathode select switches between a 90-V bus [V_{KA}(off)] and the 180-V supply. The circuit action is for the anode switches to sequentially close and at which time the selected cathodes will fire. In Fig. A4.3 are shown the anode and cathode voltages for a-segment on at anode 1 time 1 and at anode 2 time 2. Two other combinations of voltages are possible: a-segment off at time 1 and on at time 2 (a-segment on in digit 2 but off in digit 1), and a-segment on at time 1 and off at time 2. Note that since all a-segments are common, if the a-segment voltage is low (−180 V) in digit 1 at time 1, it is also low in digit 2 at time 1.

Part of the voltage doubler circuit is shown in Fig. A4.4a, where the switching transistor allows the doubler capacititor C_D to charge through the shunt diode during segment off times or during an extra scan time period every multiplex (MUX) cycle. When a cathode is selected by turning the transistor off, the capacitor's 90 V adds to the power supply's 90 V to form the V_{KK} supply for the selected segment (cathode). This action is schematically illustrated in Fig. A4.4b. When the switch is closed, all six a-cathodes are at −90 V (as a result of diode action the

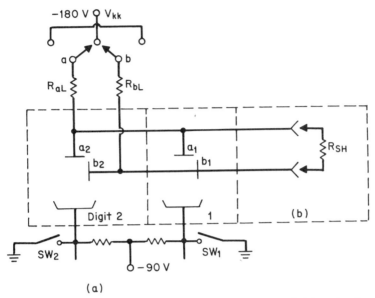

Fig. A4.1. (*a*) Schematic 2-digit, 2-segment display and drive circuits. Switch (swc) selects cathodes a and b; swl selects digit 1 and sw2 selects digit 2. (*b*) A shunt resistance may develop throught external contamination between leads or through sputtering contamination. Segment pull-down resistors are not shown.

capacitor point e is at ground); when it is open, they are all at $-90 + V_C$, where V_C is the capacitor voltage.

High-resistance cathode-to-cathode shorts may be formed in some types of tubes over time by foreign-sputtered cathode films. Alternatively, they can be formed by dirty assembly practices externally. In such a case the shunt resistor from segment a to segment b shown in Fig. A4.4b is included in the on-cathode circuit (*a*-segment, for example) as follows: The on a-segment has an open transistor switch in the circuit of Fig. A4.4b. A second separate circuit like that of Fig. A4.4a drives segment b and the transistor is conducting. Following this b-segment circuit from the 90-V supply (V_s) through the diode (or the capacitor as a sink is required), one passes to the b-cathode, then through R_{SH} to the a-cathode at voltage V_a. The shunt then effectively connects the a-segment to -90 V. The circuit of Fig. 4.4b is obtained for analysis.

At the start of the select a-segment cycle, the switch is closed. Since there is a diode in the circuit, point e is at 0 V, whereas $V_a = -90$ V. Selecting cathode a by opening the switch with b not selected starts C_D discharging into C_{tube} and R_{SH} (again $R_L < R_{SH}$ and its actual location in series with R_{SH} can be ignored). The capacitance C_{tube} is the associated cathode a capacitance and is typically in the range of 100 to 300 pF.

The tube-voltage waveshape $V_a(t)$ across C_{tube}, with and without R_{SH}, will be of interest for the analysis. Assumption of a well-regulated power supply will allow calculation of $V_a(t)$ by means of the calculation of $V_c(t)$ (Fig. 4.4c). Note that when

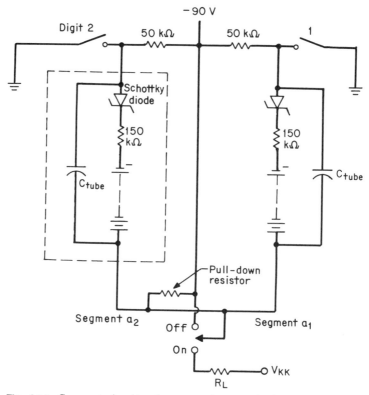

Fig. A4.2. Segment circuit redrawn to show equivalent circuit which includes shunt capacitance.

the switch opens, $V_c - 90$ V $= -180$ V. Note also that this is not the applied tube voltage which was -90 V just before the switch opened and is -90 V just afterwards.

To investigate some magnitudes of currents at the start of the cycle, the shunt current is calculated as $-V_c/R_{SH} = 90$ V$/10^6$ $\Omega \simeq 10^{-4}$ A. The tube-circuit charging current is $-V_c/R_L \simeq 90$ V$/150$ k$\Omega \simeq 50$ mA. As a first approximation the effect of R_{SH} on V_c is neglected in treating $V_c(t)$. The analysis is continued by use of Fig. A4.4c. The following set of equations is obtained:

$$i_1 = i_2 + i_3 \tag{A4.1}$$

$$V_c = +i_1 R_1 + \frac{1}{C_{\text{tube}}} \int i_3 \, dt - V_s \quad (V_s : \text{power-supply voltage}) \tag{A4.2}$$

$$V_c = +i_1 R_1 + i_2 R_2 \tag{A4.3}$$

Eliminating i_2 from Eqs. (A4.2) and (A4.3),

$$V_c = +i_1 R_1 + (i_1 - i_3)R_2 = +i_1(R_1 + R_2) - i_3 R_2 \tag{A4.4}$$

Solving Eqs. (A4.2) and (A4.4) for i_3,

$$R_2 V_c = +\frac{R_1 + R_2}{C_{\text{tube}}} \int i_3 \, dt + i_3 R_1 R_2 - V_s(R_1 + R_2) \tag{A4.5}$$

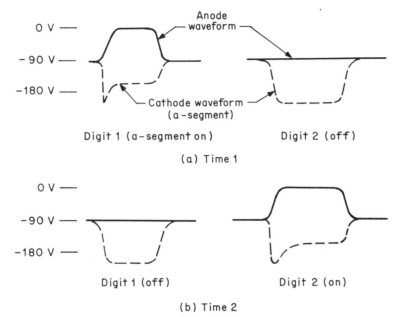

Fig. A4.3. Waveforms for a segment of the circuit of Fig. A4.1. The solid curve is the anode waveform and the dotted curve is the cathode waveform.

Let $q_3 = \int i_3 \, dt$ and $i_3 = dq/dt = \dot{q}$ in Eq. (A4.5), where the subscript of q_3 may occasionally be dropped and it is noted that $R_1 R_2/(R_1 + R_2) = R_p$. Then:

$$\dot{q}_3 + \frac{q_3}{R_p C} = -\frac{V_c}{R_1} + \frac{V_s}{R_p} \tag{A4.6}$$

Note that this equation will be useful to describe the transistor drive circuit in Sec. 4.4, under Transistor Drive, where the driving term is replaced with V_s/R_L or with the V/R_1 replaced by i_0 for the constant-current drive. It must be modified with a time-dependent V for the case under consideration.

In the first approximation of constant V, solving:

$$q_3 = A_1[1 - \exp(-t/R_p C)] + V_s C \tag{A4.7}$$

and from the boundary condition $q_3 = -V_s C$ at $t = 0$,

$$q_3 = \frac{R_p}{R_1} CV_c[1 - \exp(-t/R_p C)] + V_s C \tag{A4.8}$$

The tube voltage V_a is found as q_3/C:

$$V_a = \frac{R_p}{R_1} V_c[1 - \exp(-t/R_p C)] + V_s \tag{A4.9}$$

Before considering the behavior of V_a with and without R_2, Eq. (A4.9) will first be modified for the effects of a time-dependent V_c:

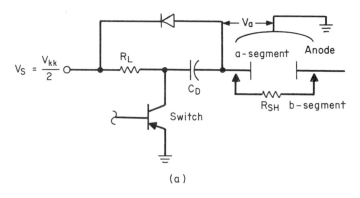

(a)

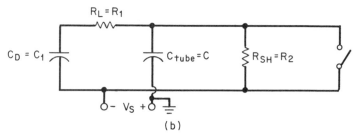

(b)

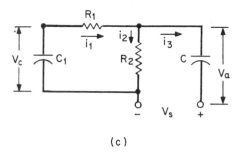

(c)

Fig. A4.4. Analysis of voltage doubler cathode driver with a-b shunt. (a) Voltage doubler circuit driving a-segment with a-b segment shunt path. (b) Redrawing of (a) with tube not fired and the tube represented by C_{tube}. This diode is simulated by tieing the switch to $-V_S$ and letting $V_C = V_S = -90$ V after the switch is opened. (c) Simplifying (b) and showing currents (switch is open).

From Fig. 4.4c note that since $\Delta q_1 = i_1 \Delta t$,

$$V_c = V_{co} - \frac{i_1 \Delta t}{C_1} \tag{A4.10}$$

is the linearized time dependence of V_c. Also noting from Eq. (A4.1) that

$$i_1 = i_3 + i_2 = \dot{q}_3 + \frac{(V_c - i_1 R_1)}{R_2} \tag{A4.11}$$

and solving for i_1,

$$i_1 = \dot{q}_3 - \frac{(V_{co} + \frac{i_1 \Delta t}{C_1} - i_1 R_1)}{R_2} \tag{A4.12}$$

and where a long time constant and nearly linear decay of V_C are assumed to avoid simultaneous differential equations:

Solving for i_1,

$$i_1 \left(1 - \frac{\Delta t}{R_2 C_1} + \frac{R_1}{R_2}\right) = \dot{q}_3 + \frac{V_{co}}{R_2} = ai_1 \tag{A4.13}$$

where $a = 1 + \frac{\Delta t}{R_2 C_1} + \frac{R_1}{R_2}$

Substituting $i_1 = \dot{q}_3/a + V_{co}/R_2 a$ in Eq. (A4.10) and replacing V_C in Eq. (A4.6), a new differential equation is devised:

$$\dot{q}_3 + \frac{\dot{q}}{R_P C(1 - \Delta t/aR_1 C_1)} = -\frac{V_{co}(1 - \Delta t/R_2 C_1 a)}{R_1(1 - \Delta t/aR_1 C_1)} + \frac{V_s}{R_p(1 - \Delta t/aR_1 C_1)} \tag{A4.14}$$

When the Δt dependence is ignored, the solution is similar to Eq. (A4.8) except that $R_p C$ is now increased to

$$R_p C(1 - \Delta t/aR_1 C_1) = RC$$

and so on, so that:

$$V_a = \frac{R_p}{R_1}(1 - \Delta t/aR_2 C_1)V_{co}[1 - \exp(-t/RC)] + V_s \tag{A4.15}$$

It is now possible to consider the cases of a shunted a-segment: There are 1 MΩ f-to-a and a-to-b shunts with a resultant R_{sh} of 0.5 MΩ. The doubler capacitor is 0.02 μF; other values are $150 - kΩ R_L$; $V_{co} = V_s = 85$ V. Tube values are: $V_K = 142$ V and $\Delta V/V_K = 15\%$ ($E_{nd} = 163$ V); $C = 120$ pF (including segment-to-segment as well as a-segment-to-ground capacitance. Equation (A4.15) then reduces to Eq. (A4.9). The results are plotted in Fig. A4.5.

It is possible to consider the behavior of the above tube with constant-current drive and shunted segments (Fig. A4.6a). However, since the voltage changes are small, its behavior differs little from the shunt-free case.

A worst-case analysis of shunt to ground (Fig. A4.6b) is more instructive. Proceeding as before,

$$V_c = \frac{q}{C} \quad \text{and} \quad i_0 = \dot{q} + \frac{V_c}{R_{sh}}$$

then the differential equation is

$$i_0 = \dot{q} + \frac{q}{R_{sh} C}$$

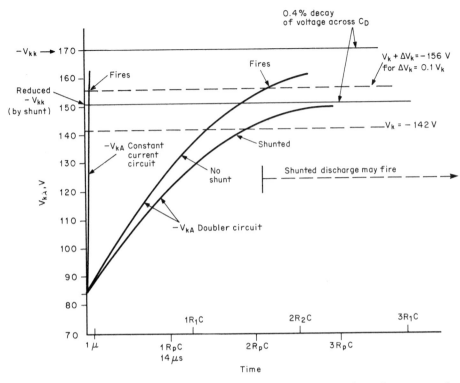

Fig. A4.5. Prefiring behavior of a GD cell with a segment-to-segment short. One segment is off; one is switched on at $t = 0$. The constant-current circuit behavior is contrasted with that of the voltage doubler circuit.

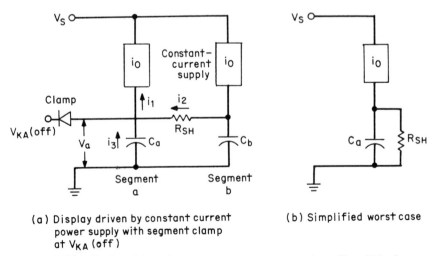

(a) Display driven by constant current power supply with segment clamp at V_{KA} (off)

(b) Simplified worst case

Fig. A4.6. The display driven by a constant-current supply suffers little from a segment-to-segment shunt. The worst case will be a shunt to ground.

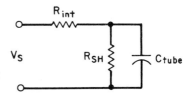

Fig. A4.7. Simplified internal-resistance limited power supply—gas discharge element prior to firing.

with the trial solution

$$q = A(1 - e^{-t/R_{sh}C})$$

and boundary conditions at $t = 0$, $q = 0$, and

$$\dot{q} = i_0$$
$$A = -i_0 R_{sh} C$$
$$V_c = i_0 R_{sh}(1 - e^{-t/R_{sh}C}).$$

When the power supply cannot supply i_0 at turn-on, the power-supply internal impedance R_i is used to limit the current (Fig. A4.7).

Suppose the power supply is a 1000-μF capacitor and the switch is such that at 3 μs it is delivering the full current i_0, then R_i is about 1 kΩ and the RC is about 0.3 μs. The tube charging waveform is shown in Fig. A4.5.

It is seen that the charging time of less than 0.4 μs is limited mainly by the power supply; the shunt is not a factor.

Note: Switching transistor voltage drops were ignored. Readers may easily include them if they wish.

APPENDIX 4.2
Parameter Estimation

Early circuit designs for gas discharge tubes may require the designer to estimate tube parameters. Table A4.1 allows quick estimates of minimum sustaining voltage $V_{KA}(\text{on})_{min}$ from the cathode fall.

The minimum striking voltage $(V_K)_{min}$ may be expressed as[4,5]

$$(V_K)_{min} = V_1 + V_2 \log \frac{1}{\gamma}$$

where the voltage is for a pd at the Paschen curve minimum. An example for a tube with a molybdenum cathode and a Ne plus ½% Ar (99.5:0.5) filling is given in Ref. 4:

$$(V_K)_{min} = 17 \text{ V} + 34.5 \text{ V} \times 2.6 = 107 \text{ V}$$

The minimum sustaining voltage $V_{KA}(\text{on})_{min}$ is calculated as

$$V_{KA}(\text{on})_{min} = V_2 + V_3 \log \frac{1}{\gamma}$$

For the example above,

$$V_{KA}(\text{on})_{min} = 17 \text{ V} + 27 \text{ V} \times 2.6 = 87 \text{ V}$$

Table A4.1.   Normal Cathode Fall

(Values in volts)

Cathode	Air	A	He	H$_2$	Hg	Ne	N$_2$	O$_2$	CO	CO$_2$	Cl
Al	229	100	140	170	245	120	180	311			
Ag	280	130	16$	216	318	150	233				
Au	285	130	165	247		158	233				
Ba		93	86				157				
Bi	272	136	137	240			210				
C				240	475				525		
Ca		93	86			86	157				
Cd	266	119	167	200		160	213				
Co	380										
Cu	370	130	177	214	447	220	208		484	460	
Fe	269	165	150	250	298	150	215	290			
Hg			142		340		226				
Ir	380										
K	180	64	59	94		68	170		484	460	
Mo					353	115					
Mg	224	119	125	153		94	188	310			
Na	200		80	185		75	178				
Ni	226	131	158	211	275	140	197				
Pb	207	124	177	223		172	210				
Pd	421										
Pt	277	131	165	276	340	152	216	364	490	475	275
Sb	269	136		252			225				
Sn	266	124		226			216				
Sr		93	86				157				
Th						125					
W					305	125					
Zn	277	119	143	184			216	354	480	410	
CsO-Cs						37					

NOTE: Values of V_{KA}(off) will be greater than but nearly equal the cathode fall. Values for Penning mixtures can be estimated by interpolation between results given for molybdenum there and the cathode material here.

SOURCE: Ref. 6.

Table A4.2.   Gas Constants

(Units are in volts)

Gas	V_1	V_2	V_3	V_4
He and He-Ne (95:5)	25	20	83	50
Ne-Ar (99.5:0.5)	17	17	35	27
He-Ar (99.5:0.5)	20	20	42	30

Table A4.3.   Value of ln (1/γ)

Cathode impinging ion	Ni	Mg	Ba	K	C
Helium			2.3	1.8	
Neon	3.8	2.2		1.5	4.3
Argon	3.0	2.6	2.0	1.5	
Hydrogen	3.0	2.1		1.5	

SOURCE: Determined from data in Ref. 6.

Referring to Fig. 4.12, it is seen that these estimates are quite good. Values of V_1, V_2, V_3, V_4, and γ may be deduced from tabulations given in Reference 4. For reference, values of ln (1/γ) calculated from Cobine[6] are given in Table A4.3, and estimated values for the other gas constants are given in Table A4.2.

Acton and Swift also show how to estimate normal glow current density:

$$j = Jp^2$$

where J is in the range of 1 to 2 $\times$ 10^6 for nobel gas Penning mixtures and p is in torr. (See also Reference 6.) For the example, they show that

$$j = 10^{-6} \times 100^2 = 10 \text{ mA/cm}^2$$

Then crudely (within 3 times either way), if the cathode is 10 cm^2 in area, 100 mA would be required to cover the segment. Reducing the pressure to reduce j has limits since the glow becomes quite diffuse at low pressures, i.e., below 40 mmHg for the mixture above.

Low-pressure tubes have large dynamic impedances and tend to require stiffer regulation of their power supplies and smaller voltage swings in multiplexed operation. In general, pressures of 60 to 150 mmHg are used with neon-argon (99.5 : 0.5) mixtures; the resulting impedances are in the 50 to 300 kΩ range.

5

The Visible Light-Emitting Diode Display

5.1 INTRODUCTION

The visible light-emitting diode (VLED) display enjoys broad acceptance. It is used in all types of products from watches and calculators to instruments, appliances, and point-of-sale (POS) applications. Other applications include channel indicators for Citizen's Band radios and televisions. It has the advantage of being easy to drive, it is inexpensive, and it has reasonable power requirements. Its disadvantages include reduced viewing angles for displays with magnifiers and possible nonuniform segments for the larger displays.

5.2 PRINCIPLES OF OPERATION

The VLED is a semiconductor diode which gives off light when forward-biased. To be visible the emitted light must have an energy in the eye sensitivity region (Chap. 3), and the semiconductor band gap must exceed the minimum energy set by the far-red eye sensitivity of about 660 nm.

This energy is:

$$Eg_{min} \geq \frac{1.24 \, eV \cdot \mu m}{0.66 \, \mu m} = 1.9 \, eV$$

It is seen that both silicon (Eg $\simeq$ 1.1 eV) and GaAs (Eg $\simeq$ 1.45 eV) are ruled out.

In the following sections VLED materials in common use and their application in VLEDs are discussed.

VLED Material

For all practical purposes, the VLED materials are red $GaAs_{0.6}P_{0.4}$ vapor epitaxially grown on germanium or gallium arsenide, and red $GaAs_{0.33}P_{0.67}$:N, yellow $GaAs_{0.15}P_{0.85}$:N, and green or yellow-green GaP:N grown on gallium phosphide substrates. Occasionally orange or amber VLED alloy compositions are produced. The gallium arsenide–gallium phosphide alloy has many of the strengths in the VLED area that silicon has in the transistor area: (1) reasonably well-understood properties and (2) a well-developed technology. Competitive materials are not as well characterized or understood.

Gallium phosphide is miscible in gallium arsenide at all compositions and the two form a well-behaved alloy system. The band gap varies from 1.45 eV for GaAs to 2.34 eV for GaP. However, the structure changes, at about a 50:50 composition, from a direct-gap material at the Γ-point (Fig. 5.1) characteristic of GaAs to an indirect one at the X-point. The quantum efficiency η (ratio of electrons in to photons out) drops sharply in this region (Fig. 5.2) for the following two reasons.

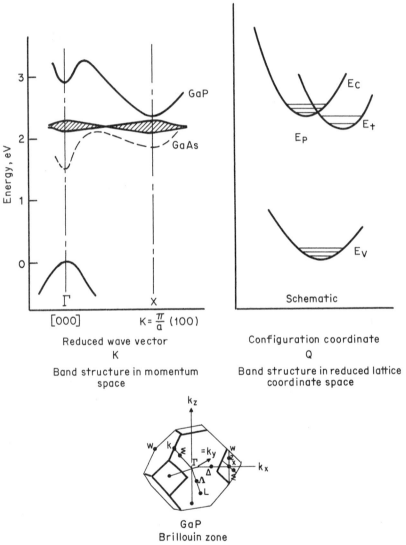

Fig. 5.1. Band structure of GaAs and GaP and Brillouin zone of GaP. The configuration coordinate diagram has been useful for understanding the spectra from localized bound states in GaP, but it has not been applied to the shallow non-localized (in K-space) case of nitrogen in GaP. *(Adapted from Ref. 3.)*

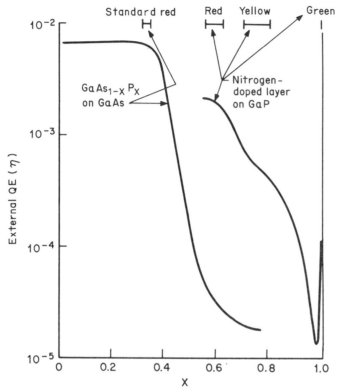

Fig. 5.2. Quantum efficiency as a function of x in $GaAs_{1-x}P_x$. *(Adapted from Refs. 7 and 8, and unpublished work.)*

The quantum-efficiency may be written as

$$\eta = \frac{1}{(1 + \Gamma_r/\Gamma_{nr})} \simeq \frac{\Gamma_{nR}}{\Gamma_R} \qquad \text{for } \eta \ll 1$$

where Γ_r is the radiative lifetime and Γ_{nr} is the nonradiative lifetime.

Since the indirect-radiative transition requires a three-body interaction to conserve momentum for the $X \rightarrow \Gamma$ transition, Γ_r here is large compared to Γ_r for the direct transition. The second factor is the loss of electrons from the direct valley to the indirect valley as the alloy shifts from GaAs to GaP. Then assuming Γ_{nr} constant and that the larger Γ_r has negligible efficiency, the second factor dominates.[1]

For this reason, the direct-gap $GaAs_xP_{1-x}$ alloys ($X \simeq .6$) are almost exclusively used for red VLEDs (standard red). It is possible to convert the indirect $GaAs_xP_{1-x}$ alloys to a pseudodirect-gap material by addition of nitrogen to form $GaAs_xP_{1-x}:N$. The improved quantum efficiency is also shown in Fig. 5.2.

In a simplified fashion one of the carrier lifetimes is related to reciprocal carrier concentration:[2]

$$\Gamma = (B_r N)^{-1}$$

where band-to-band recombination and small injection is assumed, B_r is a recombination coefficient, and N is the majority carrier concentration.

Therefore, it is necessary to optimize the material carrier concentration as a part of diode design.

Diode Operation

The forward-biased diode will emit light as some portion of the injected minority carriers radiatively recombine. Not all the radiation may be in the visible, but most of it is for well-manufactured material. A charge-carrier fate chart which is representative of nitrogen doped indirect-gap material is shown in Fig. 5.3. It is seen that there are a variety of loss mechanisms (lumped as Γ_{nr}) that must be controlled. The emitted light will be nearly proportional to the current for sufficiently high current densities (Fig. 5.4). The light intensity may be expressed as

$$L = \eta_0 i_{\text{diode}}^s \tag{5.1}$$

where i_{diode} = diode current

s = exponent (about 1.35 for GaP)

η_0 = quantum efficiency at some arbitrary current density L_0/A_d and is

$$\eta_0 = \frac{n_{\text{photon}} q}{\text{second}} \Big/ i_0 \tag{5.2}$$

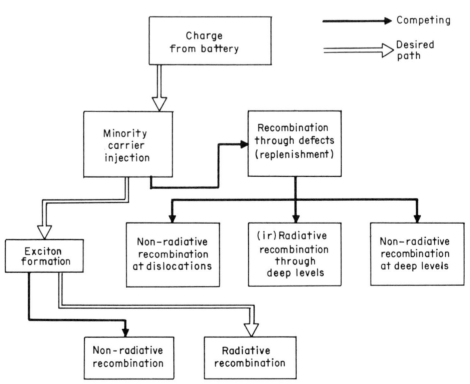

Fig. 5.3. GaP:N material—VLED charge carrier fate chart.

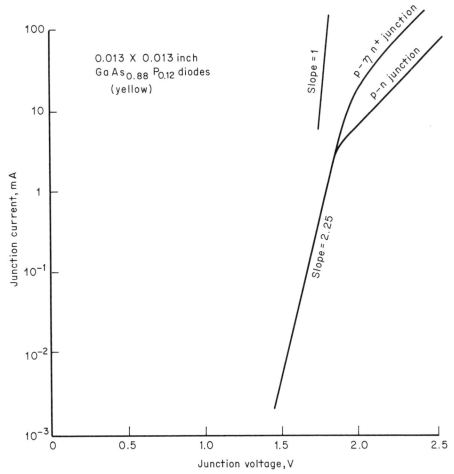

Fig. 5.4. VLED I-V characteristic. Junction diffusion current is masked by recombination-generation in the space charge region. Light output is proportional to the 1.35 power of current.

where n is the photon generation rate and

$$i_{0,\text{diode}} = i_{\text{diff.p}} + i_{\text{diff.n}} + i_{\text{recomb}} + i_{\text{shunt}} \tag{5.3}$$

where $i_{\text{diff.n}}$ and $i_{\text{diff.p}}$ = diffusion portion of current in the n– and p– layers, respectively*

i_{recomb} = space-charge recombination portion

i_{shunt} = junction leakage current and will be neglected for planar-diffused devices (all at same fixed i_0)

The typical diode area is chosen so that i_{recomb} is small at the desired operating currents (and current degradation effects of any are minimized). The departure from linearity of the light emission is commonly explained as due to saturation of

---

*By some conventions i_0 only includes this term.

competing nonradiative mechanisms.[3] It is now convenient to separate expressions (5.1) to (5.3) as:

$$L \propto i_{\text{diff}.p}\eta_p + i_{\text{diff}.n}\eta_n$$

where the nonlinear behavior is ignored and the $i_{\text{diff}.n(p)}\,\eta_{n(p)}$ is the light from the $n-(p-)$ layer.

Following Wight et al.[3] the diode bulk electroluminescent (EL) efficiency in the $n-$ plus $p-$ layer is

$$\eta_B = K(\Gamma_h\eta_n + \Gamma_e\eta_p)$$

where it is noted that surface recombination and bulk absorption is neglected and,

K = constant of proportionality

$\eta_{n,p}$ = layer quantum efficiency (may be determined by laser excitation or scanning electron microscope (SEM) excitation)

$\Gamma_{e,h}$ = injection fractional efficiency, where

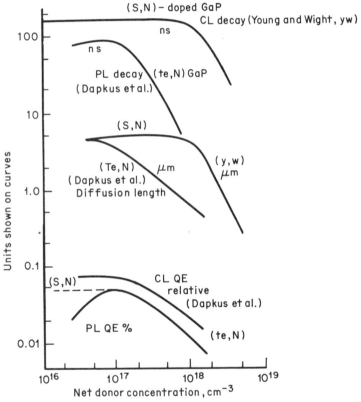

Fig. 5.5. Properties of n-type GaP:N as a function of carrier concentration. S is sulfur doping, Te is tellurium doping. CL: cathodoluminescence is luminescence observed in a scanning electron microscope with a photomultiplier tube. PL: photoluminescence is luminescence excited with a green Ar⁺ laser. [*After Ref. 9; P. D. Dapkus, W. H. Hackett, Jr., O. G. Lorimer, and G. W. Haszko, Appl. Phys. Lett., 22 (1973), p. 227; and M. L. Young and D. R. Wight, J. Phys. D., 7, (1974), p. 1824.*]

$$\Gamma_e + \Gamma_h = 1$$

and

$$\Gamma_e = \frac{L_h \sigma_n}{L_e \sigma_p + L_h \sigma_n} \qquad \text{into the p-layer}$$

and

$$\Gamma_h = \frac{L_e \sigma_p}{L_e \sigma_p + L_h \sigma_p} \qquad \text{into the n-layer}$$

The diode designer then optimizes η_B by varying the carrier concentration in the $n-$ and $p-$ layers. The variation of the quantum efficiency for GaP is given in Figs. 5.5 and 5.6, which are similar for those of yellow GaAsP:N.

Because of the nature of the recombination, the $p-$ layer has higher quantum efficiency (QE) than the $n-$ layer. The $p-$ layer to $n-$ layer QE values are given in Table 5.1 along with diffusion lengths ($L_{n,p}$) for various materials. The $L_{n,p}$ also vary with carrier concentration (for GaP see Figs. 5.6 and 5.7). The variation of L_n for GaP:N and yellow GaAsP:N can also be estimated from the variation of QE in fixed N_A GaP diodes (Fig. 5.7).

The optimum diffusion length of electrons in p-type standard red material (GaAs$_{0.6}$P$_{0.4}$) is estimated from the variation of EL with zinc doping; it apparently is optimum for N_A about 10^{18} cm^{-3}.

The results of calculations based on the above considerations for nitrogen-doped gallium phosphide are given in Table 5.2 and are compared to an experiment there.

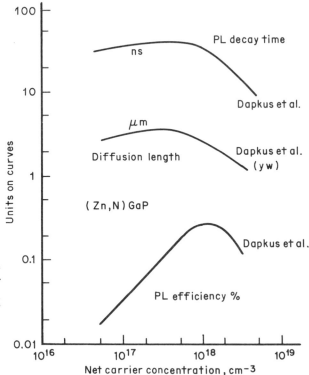

Fig. 5.6. Properties of p-type GaP:N as a function of carrier concentration for zinc-doped material. (*Based on Ref. 9 and sources for Fig. 5.5, as adapted from R. N. Bhargava, IEEE Trans. Electron Devices,* **22**, *1975, p. 697.*)

Table 5.1. Diffusion Lengths L and Ratio of *p*-type Quantum Efficiency to *n*-type Quantum Efficiency for Gallium Arsenide Phosphide Alloys and GaP:N.
All the alloys are nitrogen-doped except the standard red. The quoted values are for vapor-phase epitaxial layers and are estimated from SEM PL measurements except for the standard red, which is estimated from EBIC Measurements by J. H. Tregilas, and the second GaP:N (green), which is from the literature for liquid-phase epitaxial material.

Color	L_p (electrons in *p*-type)	L_n (holes in *n*-type)	Relative QE (η_p/η_n)
Green	0.9 μm	3 μm	3:1–6:1 (PL) measured by CL
	5–7 μm	5–7 μm	
Yellow	1.2 μm	3.8 μm	3:1–10:1
Orange	2.6 μm	3.8 μm	15:1–50:1
Standard red	0.4 μm	3–7 μm	10:1–20:1

Some estimates for vapor-phase epitaxial layers are given in Table 5.3. It is seen that there are about equal contributions from each side of the junction.

The standard red device is optimized experimentally (Fig. 5.8).

Once the doping design is accomplished, the junction depth is optimized to where it is deep enough to avoid surface recombination but shallow enough to avoid significant light absorption. About 2 μm has been found optimum for direct-gap (standard red) GaAsP. Absorption does not seem to be critical for indirect materials (Fig. 5.9) and surface recombination, doping profile, and nitrogen layer thickness are optimized (deeper junctions are more costly).

An optimization procedure for a yellow GaAsP device is shown in Fig. 5.10.

An additional factor involved in high EL is efficient light extraction. It is accomplished by use of an antireflective coating such as the SiO_2 or Si_2N_3 diffusion mask and/or the plastic conformal coating. Additionally, for indirect-gap materials,

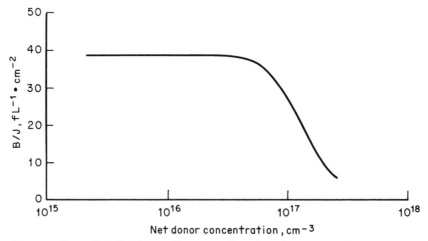

Fig. 5.7. Green GaP:N VLED efficacy as a function of *n*-type carrier concentration. The *p*− layer concentration is fixed and is set by the diffusion process. *(After Ref. 4.)*

Table 5.2 GaP:N Electroluminescence Efficiency
Calculations are compared to experiment (Exp.)

Doping		CL efficiency		$\dfrac{L_e}{L_n}$	$\dfrac{\sigma_p}{\sigma_n}$	Injection efficiency		EL efficiency × $10^{-3}\%$	
p-type	n-type	R_p	R_n			τ_e	τ_h	Pred.	Exp.
p^+	n	0.78:0.5		0.40	4.2	0.37:0.63		0.60	0.5
p^+, N	n	12:1.0		0.40	4.2	0.37:0.63		5.0	4.0
p, N	n^+	16:0.40		0.80	0.36	0.78:0.22		13	15
p^+	n, N	1.0:9.1		0.55	3.5	0.34:0.66		6.4	8.0

SOURCE: Ref. 3, WBTY.

Table 5.3. Estimates of EL for Vapor-Phase Epitaxial Layers

$\dfrac{\sigma_p}{\sigma_h}$	$\dfrac{L_e}{L_h}$	$\dfrac{p}{\tau_e}$	$\dfrac{n}{\tau_h}$	$\dfrac{N_p}{N_n}$	Relative EL $(p + n)$
4.2	0.4	0.373	0.623	3	$1.11 + 0.623 = 1.74$
10	0.33	0.23	0.77	3	$0.69 + 0.77 = 1.44$

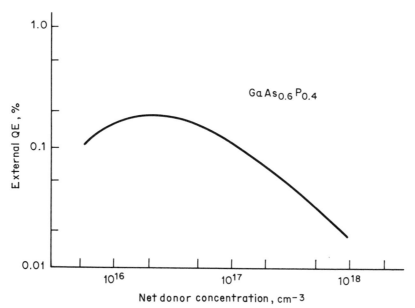

Fig. 5.8. Optimization of red VLED by donor concentration change. *(After Ref. 10.)*

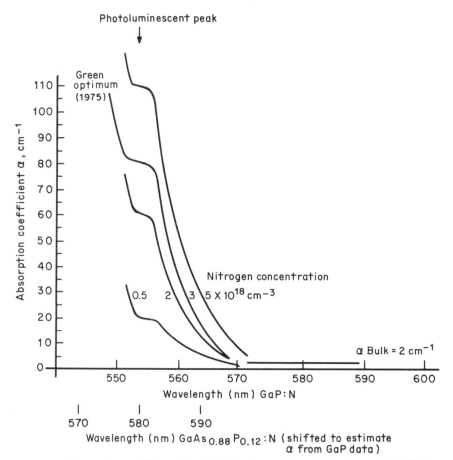

Fig. 5.9. Absorption coefficient for GaP:N estimated from Refs. 11 and 12. Yellow VLED GaAsP absorption may also be roughly estimated from these curves by shift of the wavelength scale.

a reflecting back contact is used and a means for collecting the light leaving the die sides is provided.

Color

For colors other than red, the peak wavelength may be used along with Table 3.5 to specify the color. However, the indirect-gap materials have significant radiation in the red, orange, or yellow regions for orange, yellow, or green devices.

Stringfellow and Kerps[4] have suggested use of the centroid to specify the wavelength for a representative device. This procedure is summarized for yellow-green GaP devices in Fig. 5.11.

The wider emission spectrum requires the designer to use more care with filter design so as to not throw away much of the light (Chap. 3 and Fig. 5.12).

VLED Brightness

VLED brightness will depend on the eye sensitivity to its emitted color. Ideally the brightness is calculated by multiplying the emitted power in each wavelength

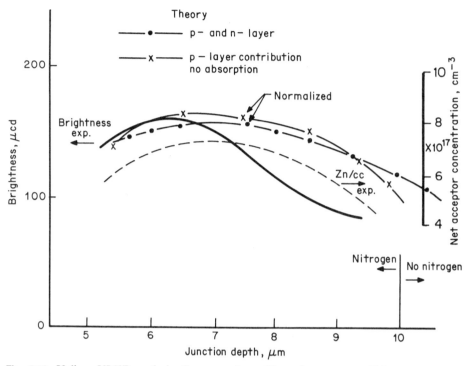

Fig. 5.10. Yellow VLED optimization procedure. Here the acceptor (Z_n) concentration varies with diffusion depth. The calculated brightnesses for the deeper junctions are higher than the experimental values primarily because there is no nitrogen (QE $\div$ 10) beyond 10 μm. *(Experimental data from Ref. 13.)*

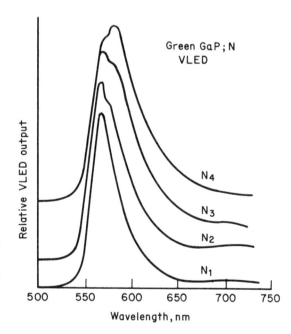

Fig. 5.11. As the nitrogen is increased from N_1 to N_4, the amount of red and yellow light increases relative to the green (peak). Rather than specifying peak wavelength for GaP diode emission, some manufacturers report the centroid wavelength; e.g., $N_1 = 573$ nm; $N_2 = 576$ nm; $N_3 = 578$ nm; $N_4 = 582$ nm. *(From Ref. 4.)*

interval by the eye response (Fig. 5.13) and summing. In practice, the standard red diodes have sharp spectra; the multicolor GaAsP and GaP may also be filtered to a near sharp line spectra (Fig. 5.12). Therefore, the peak wavelength can be used to determine brightness without integrating.

Figure 5.13 shows brightness calculated in the first fashion for various materials. It is seen that nitrogen-doped devices are almost as bright as the standard red and that the loss of output efficiency for the yellow and green devices is offset by the increased eye sensitivity.

5.3 THE VLED DISPLAY

Display Configurations

Various techniques are used to provide different VLED display sizes. The ¼ cm (0.1 in) and ½ cm (0.2 in) 7-segment displays are commonly a monolithic bar with magnifier (Figs. 5.14a and b). The ¾ cm (0.3 in) and larger 7-segment display usually is one or two individual chips per segment mounted in a light pipe or reflector package (Figs. 5.15a and b). Multidigit assemblies are also commercially available. The alphanumeric may be assembled as shown in Fig. 5.16.

A variety of magnifier designs are in use, which vary from a single-bubble lens to a bubble lens and cylindrical lens combination.[5] While most lens designs maintain the display brightness after magnification, it will be at the expense of the viewing angle. Off-axis viewing will also lead to character distortion. If a standard design does not meet the system requirements, the designer should plan to work closely with the vendor to select the optimum lens.

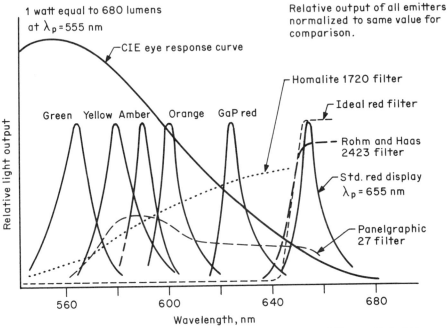

Fig. 5.12. Spectral characteristics of VLED emitters and filters. *(Ref. 17.)*

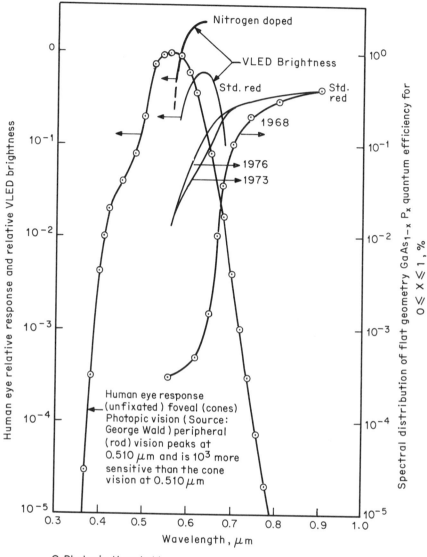

⊙ Photopic threshold

1.85 X 10⁻¹⁴ g · cal/cm²/s (at 0.510 μm, Δt = 10⁻³ s)

1.85 X 10⁻¹⁶ g · cal/cm²/s (at 0.510 μm, Δt = 1 s)

Fig. 5.13. Comparative plots of the human eye $GaAs_{1-x}P_x$ emitter spectral response, and VLED brightness. VLED improvement is based on M. G. Craford, *Proc. of the Soc. for Inf. Disp.,* **18** (1977), p. 160. The nitrogen-doped VLED brightness relative to the standard red brightness is based on collecting the fraction radiated out the sides of the VLED. Many display packages lose this radiation; then the two are comparable.

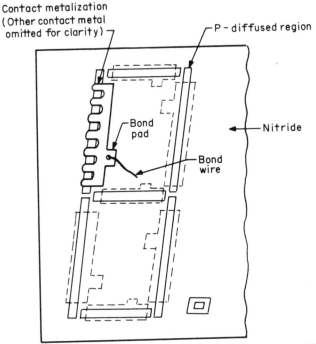

Fig. 5.14a. A monolithic 7-segment display. This pattern will be repeated four times for a watch display.

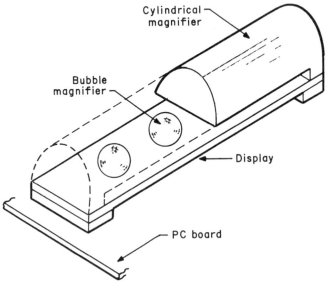

Fig. 5.14b. Magnifier assembly shown placed over display bar assembly.

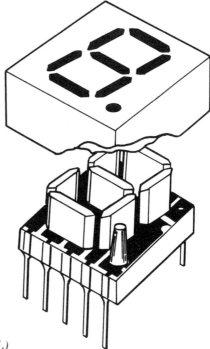

Fig. 5.15a. Light pipe package *(Ref. 15.)*

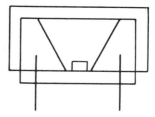

Fig. 5.15b. Reflector package *(Ref. 15.)*   The VLED die is mounted inside an open cavity with a diffusing cover

Fabrication

Figure 5.17 shows the various VLED processing steps. The substrate slice is sawed from a grown single crystal and then polished. The device material is epitaxially grown onto the substrate, usually by vapor-phase deposition in a large reactor similar to that used for silicon epitaxy. The device is then zinc-diffused to form the p− layer and metallized to form the contacts. Finally, it is assembled and packaged to form the finished display.

No special burn-in steps are required but 48-hour burn-in (usually at special request) aids stabilization.

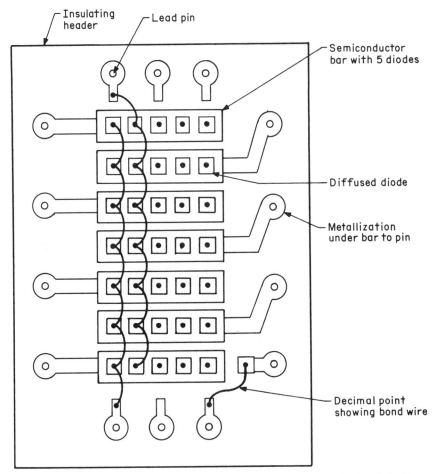

Fig. 5.16. 5×7 alphanumeric display. In this configuration the display is built up from 7 individual bars of 5 diodes each. The complete assembly may be encapsulated in plastic.

Reliability and Environmental Conditions

Early device failures were characterized by two primary failure mechanisms: brightness degradation and bond failure. The degradation problem has been largely eliminated, for devices operated within specification limits, by new processes and careful design procedures.

Small brightness drops that occur in the first 48 hours may be eliminated by initial burn-in procedures. The pulled-bond problem has been sharply reduced by means of compliant coatings with closer expansion match to the device.

Past degradation of green gallium phosphide devices is thought to be partly due to generation of dislocations as a result of the energy from nonradiative recombination at defects.[6] Reliability data is generally available from most manufacturers on request.

The effect of temperature on VLEDs is shown in Figs. 5.18 and 5.19.

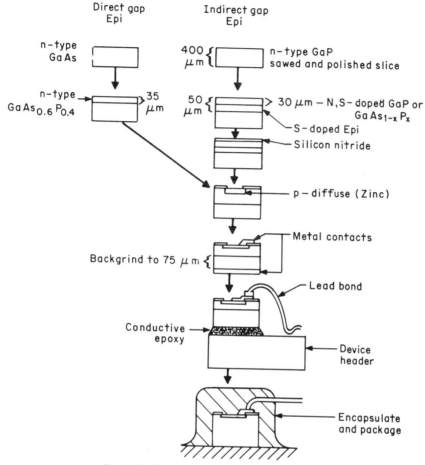

Fig. 5.17. Representative VLED processing.

Special Considerations

The plastic packages of VLED displays require special provisions for cleaning; use of only Freon TF, isopropanol, or water is recommended (at room temperature and for short times, such as 2 minutes maximum). These solvents are also the only ones to be used for flux removal following assembly. During soldering or other processing operations, temperature limits must not be exceeded as follows: circuit board assemblies should be kept below 230°C for less than 5 s. The display package or lens should be kept below 70°C. Lens assemblies should only have the edge board immersed in solders or cleaners.

5.4 DRIVE CIRCUIT REQUIREMENTS

General Driving Requirements

Except for a small offset current the VLED output is nearly proportional to current (Sec. 5.2). A constant-current driver is used (Fig. 5.20). Expected operating volt-

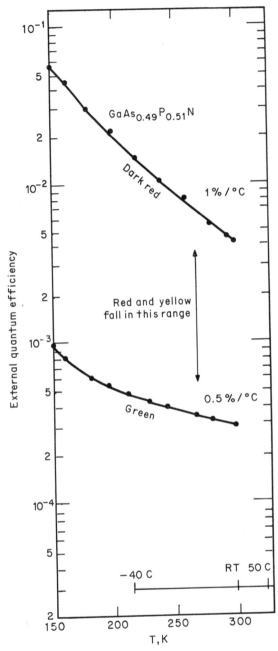

Fig. 5.18. Variation in light output of $GaAs_{1-x}P_x:N$ ($0.51 \leq x \leq 0$) diodes with temperature. *(After Ref. 14.)*

ages are: GaP, 2.2 V; yellow and amber GaAsP, 2.2 V; and red and standard red GaAsP, 2 V.

Both common-anode and common-cathode configurations are available for most displays and allow flexibility in system interfacing. Alphanumeric displays are arranged in an x-y column-row configuration to allow multiplexed operation.

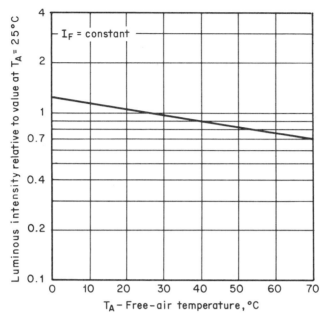

Fig. 5.19. Temperature dependence of standard red VLED. *(After Ref. 15.)*

Typical IC drivers are given in Table 5.4.

A typical 7-segment drive is shown in Fig. 5.21 and a alphanumeric circuit is shown in Fig. 5.22.

Multiplexed Displays

Two factors are involved in the use of multiplexed displays: increased quantum efficiency for drive currents above certain levels (about 5 mA for the green) and maintenance of the average segment current without exceeding the maximum allowable drive current. The second requirement will be clearly specified on the data sheet. While the first is generally ignored (brighter displays are a design

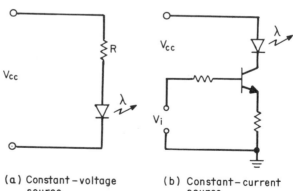

Fig. 5.20. VLED drive circuits.

(a) Constant−voltage source

(b) Constant−current source

Table 5.4. Display Drivers for Commercial Temperature Range

Display type	Description	Input compatibility	Power supplies	Drivers per package	Device type	Package type	Additional features
	Segment drivers	MOS	10 V	4	SN75491	N	50-mA source/sink capability
			20 V	4	SN75491A	N	
			Variable from 3.2 to 8.8 V	4	SN75493	N	50-mA regulated source capability Display blanking provisions
			10 V	6	SN75492	N	250-mA sink capability
			20 V	6	SN75492A	N	
LED displays		MOS	Variable from 3.2 to 8.8 V	6	SN75494	N	250-mA sink capability Display blanking provisions
	Digit drivers	MOS, TTL	Variable from 2.7 to 6.6 V	7	SN75497	N	100-mA sink capability Input threshold: 2.7 V max Low-voltage saturating outputs (0.4 V maximum)
		MOS, TTL	Variable from 2.7 to 6.6 V	9	SN75498	N	100-mA sink capability Input threshold: 2.7 V max

SOURCE: Ref. 15.

118

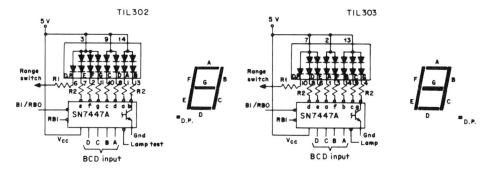

NOTES: A. R1 and R2 are selected for desired brightness.
  B. SN74L47 may be used in place of SN7447A in applications where segment forward current will not exceed 20 mA.

FUNCTION TABLE
SN7447A

DECIMAL ON FUNCTION	INPUTS					BI/RBO†	SEGMENTS							NOTE
	LT	RBI	D	C	B	A		a	b	c	d	e	f	g	
0	H	H	L	L	L	L	H	ON	ON	ON	ON	ON	ON	OFF	1
1	H	X	L	L	L	H	H	OFF	ON	ON	OFF	OFF	OFF	OFF	1
2	H	X	L	L	H	L	H	ON	ON	OFF	ON	ON	OFF	ON	1
3	H	X	L	L	H	H	H	ON	ON	ON	ON	OFF	OFF	ON	1
4	H	X	L	H	L	L	H	OFF	ON	ON	OFF	OFF	ON	ON	1
5	H	X	L	H	L	H	H	ON	OFF	ON	ON	OFF	ON	ON	1
6	H	X	L	H	H	L	H	OFF	OFF	ON	ON	ON	ON	ON	1
7	H	X	L	H	H	H	H	ON	ON	ON	OFF	OFF	OFF	OFF	1
8	H	X	H	L	L	L	H	ON	ON	ON	ON	ON	ON	ON	1
9	H	X	H	L	L	H	H	ON	ON	ON	OFF	OFF	ON	ON	1
10	H	X	H	L	H	L	H	OFF	OFF	OFF	ON	ON	OFF	ON	1
11	H	X	H	L	H	H	H	OFF	OFF	ON	ON	OFF	OFF	ON	1
12	H	X	H	H	L	L	H	OFF	ON	OFF	OFF	OFF	ON	ON	1
13	H	X	H	H	L	H	H	ON	OFF	OFF	ON	OFF	ON	ON	1
14	H	X	H	H	H	L	H	OFF	OFF	ON	ON	ON	ON	ON	1
15	H	X	H	H	H	H	H	OFF	OFF	OFF	OFF	OFF	OFF	OFF	1
BI	X	X	X	X	X	X	L	OFF	OFF	OFF	OFF	OFF	OFF	OFF	2
RBI	H	L	L	L	L	L	L	OFF	OFF	OFF	OFF	OFF	OFF	OFF	3
LT	L	X	X	X	X	X	H	ON	ON	ON	ON	ON	ON	ON	4

H = high level (logic 1 in positive logic), L = loq level (logic O in positive logic), X = irrelevant.

†BI/RBO is wire-AND logic serving as blanking input (BI) and/or ripple-blanking output (RBO).

NOTES: 1. The blanking input (BI) must be open or held at a high logic level when output functions 0 through 15 are desired. The ripple-blanking input (RBI) must be open or high if blanking of a decimal zero is not desired.
  2. When a low logic level is applied directly to the blanking input (BI), all segment outputs are off regardless of any other input.
  3. When the ripple-blanking input (RBI) and inputs A, B, C, and D are at a low logic level with the lamp test input high, all segment outputs are off and the ripple-blanking output (RBO) of the decoder goes to a low level (response condition).
  4. When the blanking input/ripple blanking output (BI/RBO) is open or held high and a low is applied to the lamp-test input, all segments are illuminated.

NUMERICAL DESIGNATIONS–RESULTANT DISPLAYS
RECOMMENDED DECODE / DRIVE WITH BCD INPUTS

Fig. 5.21. 7-segment drive circuit. *(After Ref. 15.)*

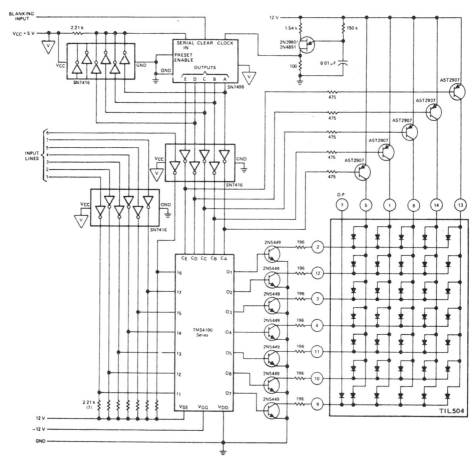

Fig. 5.22. α-N drive circuit. *(After Ref. 15.)*

margin), it may be taken into account by using curves such as Fig. 5.23 to estimate the increased brighteners at the new peak current over the data sheet values at the specified current.

REFERENCES

1. For a more detailed development see B. W. Hakki, *J. Appl. Phys.,* **42** (1971), p. 4981.
2. H. Kressel and H. Nelson, "Properties and Applications of III-V Compound Films Deposited by Liquid Phase Epitaxy," in *Physics of Thin Films,* vol. 7, Academic Press, New York (1973), p. 115; and H. Kressel and J. K. Butler, *Semiconductor Lasers and Heterojunction LED's,* Academic Press, New York (1977), chap. 1.
3. D. R. Wight, J. C. H. Birbeck, J. W. A. Trussler, and M. L. Young, *J. Phys. D,* **6** (1973), WBTY, p. 1622; and P. D. Dapkus, W. H. Hackett, Jr., O. G. Lorimor, and R. Z. Backrach, *H. Appl. Phys.,* **45** (1974), p. 4920. Also A. R. Reisinger and E. G. Bylander (private communication).
4. G. B. Stringfellow and D. Kerps, *Solid-State Electron.,* **18** (1975), p. 1019.
5. P. Jeung and J. Vebling, *Proc. of the Soc. for Inf. Disp.,* **17** (1976), p. 138.

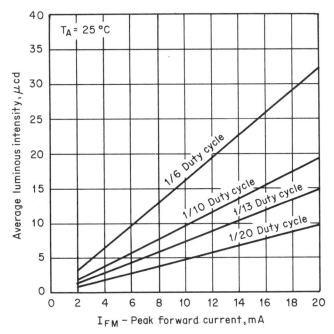

Fig. 5.23. Average luminous intensity (per segment) vs. peak forward current. *(After Ref. 15.)*

6. W. A. Brantley, O. G. Lorimor, P. D. Dapkus, S. E. Haszko, and R. H. Saul, *J. Appl. Phys.*, **46** (1975), p. 2629.
7. M. G. Craford and N. Holonyak, Jr., "The Nitrogen Isoelectronic Trap . . . ," in *Optical Properties of Solids: New Developments,* B. O. Seraphin (ed.), North Holland, Amsterdam (1975), chap. 5 (band structure); and M. L. Cohen and T. K. Bergstresser, *Phys. Rev.,* **141** (1966), p. 789 (B.Z.).
8. R. A. Logan, P. J. Dean, H. G. White, and W. Wiegman, *J. Appl. Phys.,* **42** (1971), p. 2328.
9. Adapted from Ref. 3 and M. L. Young and D. R. Wight, *J. Phys. D,* **7** (1974), p. 1824.
10. M. G. Craford, "Properties and Electroluminescence of the $GaAs_{1-x}P_x$ Ternary System," *Prog. Solid State Chem.,* **8** (1973), p. 127.
11. Adapted from R. Z. Bachrach and O. G. Lorimer, *Phys. Rev. B,* **7** (1973), p. 700.
12. R. A. Logan, H. G. White, and W. Wiegmann, *Solid State Electron.,* **14** (1971), p. 55.
13. C. L. White (private communication).
14. M. G. Craford, D. L. Keune, W. D. Groves, and A. H. Herzog, *J. Electron Mater.,* **2** (1973), p. 137.
15. Texas Instruments, Inc.
16. A. A. Bergh and P. J. Dean, *Light-Emitting Diodes,* Clarendon Press, Oxford (1976).
17. G. Novis and C. Smith, *Optoelectronics Applications,* Texas Instruments, Inc., Dallas (1977).

APPENDIX 5.1
Micropressor Interfacing

Micropressor interfacing is straightforward. For numeric or hexadecimal display, a display subsystem is first designed which uses a 4-bit to 7-segment (or hex) converter. Alternatively the display will have a built-in converter. For a simple display an address decoder is used to select a particular display during data strobing, and data latches are provided for data and decimal inputs. Programming involves storing data in the accumulator and shifting out to the specified address. A similar procedure is followed for an input-output interface except that strobed input buffers are used.[17]

6

Vacuum Fluorescent Display

6.1 INTRODUCTION

The vacuum fluorescent (VF) display is widely used in clocks and calculators. It has the advantages of a blue-green emission, of a wide viewing angle, and of relatively low-voltage and low-power operation. It has the disadvantage of low contrast in normal ambients and requires relatively transparent bandpass filters, which can lead to distracting reflections from the display cover glass.

6.2 PRINCIPLES OF OPERATION

The VF tube is a display version of the original electric tuning eye. It is a triode vacuum tube with phosphor-coated anodes. The following sections describe the operation.

Tube Display Configuration

A typical tube geometry is shown in Figs. 6.1 and 6.2. The filament is typically 18 μm (0.0003 in) in diameter and is electrophoretically coated to 25 μm (0.001 in)

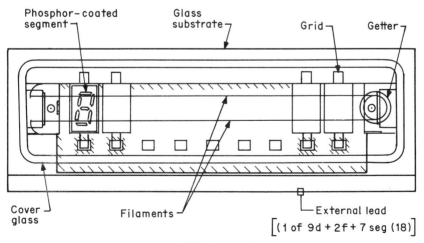

Fig. 6.1. VF display schematic.

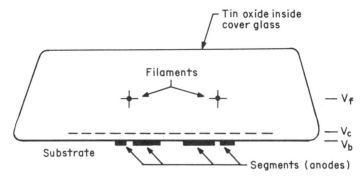

Fig. 6.2. Schematic cross section of a VF tube. A representative device would be about 0.4 inch across and 0.12 inch high.

diameter. After forming, the coating has an emissivity of 0.2, which leads to very-low-power filament operation at a temperature of about 700°C. The operating temperature is selected to be high enough to avoid poisoning and low enough to avoid barium depletion. The grid is usually etched stainless steel and it has an optical transparency of about 90%. The anodes are screened silver with a graphite buffer layer over which the zinc oxide phosphor is electrophoretically coated. Screened leads are brought through the seal ring to epoxied leads.

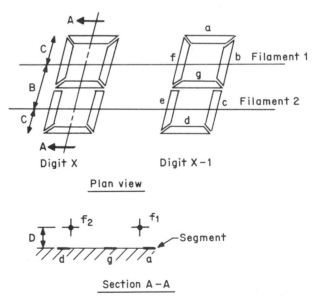

Fig. 6.3. Geometry of the VF tube showing the filament placement (grid omitted). The dimensions B and C are in the ratio of 2 to 1. The filament height (dimension D) will be slightly less than dimension C (or $B/2$).

Tube Operation

The phosphor brightness in a VF tube is linear with current; therefore, one object is to obtain as much current as possible to the anode at as low a voltage as possible. Since vacuum tube currents vary as $V^{3/2}/d^2$ (V—voltage, d—spacing), this requirement necessitates close spacing (small d). Close spacing can lead to uniformity problems; therefore a compromise spacing is chosen. Where two filaments are used (Fig. 6.3), the filament digit spacing is $\frac{1}{4}$, $\frac{1}{2}$, $\frac{1}{4}$, as shown, since both filaments supply equal currents to the g-segment. This geometry then sets an effective d (to be defined later) of roughly $\frac{1}{8}$ to $\frac{1}{4}$ mm for the smaller displays. The filament diameter is then chosen so that its emission is space charge limited at the design temperature (Fig. 6.4).

Appendix A6.1 shows how the grid and segment currents are determined. The appendix also shows how segment current depends on grid and segment voltages. Grid potentials greater than the natural grid potential (about $\frac{1}{2}V_s$ for an equispaced

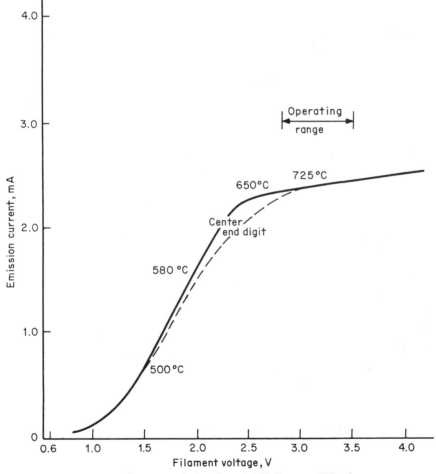

Fig. 6.4. Filament emission for a 3.0-V filament VF tube.

grid) can be wasteful of power at slight increases in brightness. However, circuit and contrast requirements usually dictate $V_g = V_s$.

Phosphor Principles

The phosphor is a zinc-doped zinc oxide or P15 type originally developed for flying spot scanner operation. It is unique in that its turn-on voltage is about 3 V above the potential for initial current flow (known as the contact potential difference, or CPD).

Its spectra is shown with the eye response and effective response in Fig. 6.5. The phosphor is in the form of grains 1 to 10 μm in diameter. The impinging electron energy is converted to electron-hole pairs in the first 100 nm[1] and to heat thereafter. Light output is linear with current, but not voltage (Fig. 6.6). There is an average of one electron-hole pair formed per primary electron at the small plateau (above 11 V + 2 V = 13 V) and two pairs are formed at 24 V. The holes recombine at centers which have energies less than the band gap energy of 3.26 eV, and the light is only weakly absorbed. About 90% is returned to the observer. The quantum efficiency at 24 V can be as high as 30%.

Typical grid cutoff voltages are in the range of -2 to -5 V. The brightness will be reduced to negligible amounts at a V_g of around $+1$ to $+2$ V. There may be some phosphor lag (Fig. 6.7), which will limit the minimum drive-pulse width.

Table 6.1 compares some 9-digit portable calculator displays.

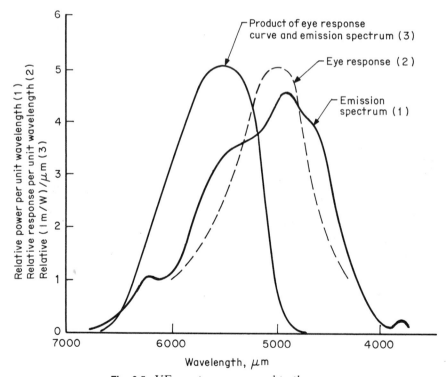

Fig. 6.5. VF spectrum compared to the eye response.

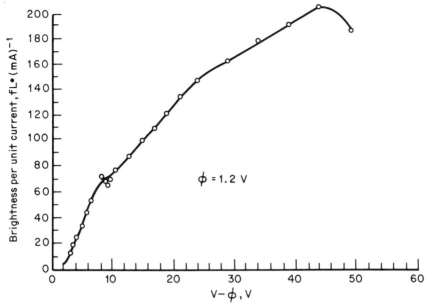

Fig. 6.6. VF brightness-voltage curve (10% duty cycle).

6.3 THE VF DISPLAY

The VF display is manufactured in 10- through 40-character, 5 × 7 format, and 7-segment and 14-segment versions, among others. It uses much of the same manufacturing technology as described earlier for the gas discharge display and which will be summarized in the next section.

Fabrication Methods

The segment and overlay screening or alternatively the thin-film processing is the same as for the gas discharge display. Next the phosphor is deposited in an electrophoretic cell by interconnecting all the segments in a jig and applying a voltage between them and a counter electrode. The phosphor is held in suspension

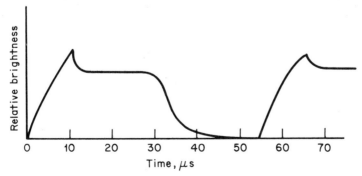

Fig. 6.7. Rise and fall times for an early VF tube.

Table 6.1. 9-digit, 0.2-in-high Character Data Sheet Comparison

Parameter	Tube: ISE FG	FUTABA				NEC			
	95A	9ST-12	9ST-08	9ST-02	LS8118	LD8091	LD8106	LD8085	LD8133
Fil:									
V, V	3 ± 3	3	3.2	2.75	3.0	2.8	3.0	2.7	2.2
I, mA	22 ± 2	22	21	40	23	35	23	58	23
W, mW	66	66	67	100	69	98	69	157	51
Grid:									
V_f, V	24	22	30	30	24	24	24	36	22
I, mA	2	2	2.5	2.5	2.0	3	1.5	4	1
V_{co}, V	−4	−4			−3	−5	−6		−2
Segment:									
V, V	24	22	30	30	24	24	24	36	22
I, mA	1.5	.9	1.2	1.2	1.0	1.5	.6	1.5	.4
V_{co}, V		−4							
Brightness, fL	70/100	70/150	50/150	50/150	50	80	80	RD	70
Form factor	Flat 75 × 26	Flat	Flat	Flat	Flat	RD	RD	RD	RD
$L \times W$ (from data sheet), in	3 × 1.04	3.2 × 0.8	2.8 × 0.8	2.8 × 0.8	2.64 × 0.84	3.3	3.1	3.6	2.0
t (measured), mils	280	280	280	280	280	740	520	740	400
Duty cycle	1/10	1/12.5		1/12	1/12	1/12			1/12
Power consumption									
P_t/digit		12 mW							
Measure $L \times W$, in	2.55 × .85	2.55 × .80			2.58 × 0.80				1.95
t, mils	270	280							400
Dead length, mils		770							720

in a dielectric liquid in the cell and deposits compactly and uniformly onto the segments. Previously etched grids and filament posts are next assembled to the substrate by means of conducting paste. They are fired and reinforced with overlay paste. The filaments are coated by pulling 18-μm (0.0003-in) tungsten through a sleeve filled with triple carbonate binder-polar liquid mixture. When a voltage is applied between the wire and the sleeve, a coating is formed which dries on the way to the take-up spool.

The filament wire is spot-welded to the filament post and spring and a getter ring is also welded in place. The cover glass and exhaust tubulation is then fired to the substrate. Following leak checking, the assembly is evacuated and baked, and the getter is preflashed. A voltage is applied to the filament to thermally activate it.

The tube is tipped off and the getter is flashed. Contacts are epoxied to the substrate, baked, and sheared. Then the tube is aged on an aging rack for about 30 minutes and tested there. The aging is primarily to clean up the electrodes and stabilize the filament and phosphor.

Reliability and Environmental Considerations

The primary failure mechanisms for the VF tube are brightness degradation due to phosphor aging by ion bombardment and filament emission drop due to contamination (especially at the ends). Initial zeros in calculators are particularly susceptible to this problem since they degrade faster through more use than the other digits. There is also a long-term emission drop due to barium loss. Filament breakage is not a particularly severe problem; although rare, a break of one filament in a two-filament tube will result in only about a 20 to 50% brightness drop of those segments near the broken filament. A broken filament can short some digits but again this is a rare failure.

Phosphor life to half brightness is about 10,000 to 12,000 hours and filaments operated at 700°C have an estimated 5000 to 7000 hours before emission limiting reduces brightness by one-half again on the ends. Consumer calculators used about 300 hours a year would have a life to half brightness of 2 to 3 years. Automobile clocks used 500 hours a year would experience half brightness in about two years with the dimming of the reading on the leading *one* offset by its decreased use.

These are not catastrophic failures and probably are not objectionable. New processes may extend operating life times so that appliance and clock applications become more feasible.

Large shocks due to dropping should be avoided to prevent phosphor dislodge or structural failures. The effect of temperatures on brightness is shown in Fig. 6.8. Manufacturers typically do not re-rate the filament at low or high temperatures because the low conductivity of the wire allows little heat gain or loss by conduction.

Special Considerations

Antimoisture agents may be required to prevent leakage and crosstalk.

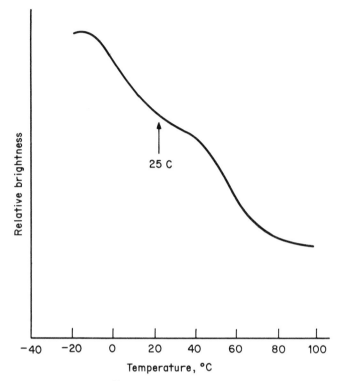

Fig. 6.8. Effect of temperature on the P15 phosphor.

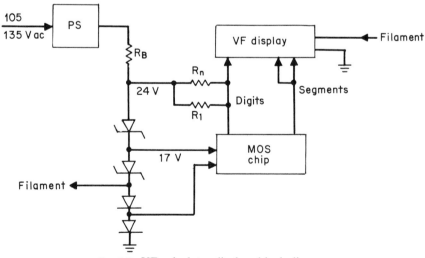

Fig. 6.9. VF calculator display: block diagram.

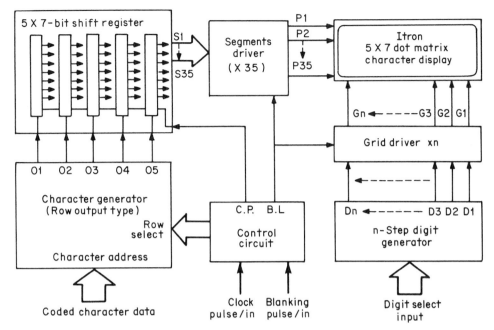

Fig. 6.10. Block diagram of the 5 × 7 drive circuit. *(From Ref. 3.)*

6.4 DRIVE CIRCUITS

The driving circuits are a cross between the gas discharge and VLED circuits. To prevent crosstalk, care must be used to avoid having even a few megohms resistance between the common power supply connections and adjacent segments.

The very-low-power filaments in multiplexed tubes can be pulse-heated as long as the rms pulse power equals the heater power requirement and the pulse comes at the end of the MUX cycle blanking period (to avoid grid-cathode debiasing effects).

There seem to be few published VF calculator circuits; a transformerless version is shown in Fig. 6.9. Better practice, of course, is to use an isolation transformer; 6.3 V ac filament VF tubes are available for such applications. A block diagram of a drive circuit for an alphanumeric VF display is shown in Fig. 6.10.

REFERENCES

1. E. G. Bylander, *J. Appl. Phys.,* **49** (1978), p. 1188.
2. Karl R. Spangenberg, *Vacuum Tubes,* McGraw-Hill, New York (1948).
3. ISE Electronics.

APPENDIX 6.1
Calculation of Tube Currents

It is easy to show experimentally that segment current is mostly line-of-sight electrons by using narrower and narrower grids (Fig. A6.1). At the point where the

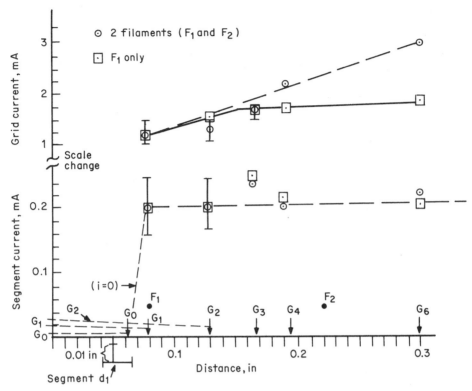

Fig. A6.1. Segment and grid currents for grids G_0 and G_6, which extend from 0 to the ordinate distance shown. Also shown are the relative filament and segment locations vertically above and below the axis.

grid is not in the segment-to-filament line of sight, there is no segment current. (No current flows to a phosphor segment not covered by a grid, since in this case the phosphor charges to the filament potential.) Also, it is verified experimentally that the majority of d-segment current is from the nearest filament only (lower curve of Fig. A6.1).

TRIODE THEORY APPLICATION

The current density J in a plane diode is

$$J = \frac{A V^{3/2}}{d^2}$$

where $A = 2.335 \times 10^{-6}\ A V^{-3/2}$
V = a voltage nearly equal to the filament-plate difference
d = a distance nearly equal to the filament-plate separation

If the geometry of Fig. A6.2 is set up, the current flowing in a straight line in the segment space $i(x)\ dx$ becomes

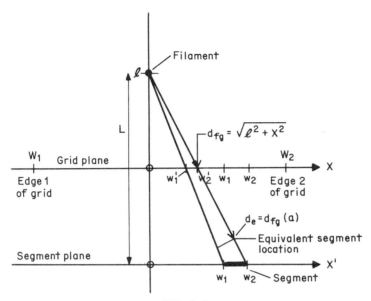

Fig. A6.2. VF triode geometry.

$$i(x) \, dx = J(x) \, dx$$

where dx = an incremental segment width projected onto the grid plane
y = segment average length and the d and V above are triode equivalent distances and voltages (Table A6.1).

The segment current is calculated as:

$$i_{\text{seg}} = \int_{w_1}^{w_2} i(x) \, t \, dx = \int_{w_1}^{w_2} J(x) yt \, dx = yt \int_{w_1}^{w_2} \frac{A V^{3/2}}{a(x)(l^2 + x^2)} \, dx$$

Table A6.1   Triode Equivalent Voltage and Distance

Conventional triode theory[2] yields the result (for $t_{\text{opt}} > 0.9$) that the equivalent spacing, d_e, is:

$$d_e \simeq d_{fg}\left(1 + \frac{d_{fg} + d_{gs}}{\mu d_{fg}}\right)$$

where d_{fg} = filament-to-grid distance along the line of sight
d_{gs} = grid-to-segment distance
μ = amplification factor for the segment-grid-filamentary cathode triode.

$$V_e = (V_g - \Psi_s) + \frac{V_s - \Psi_s}{\mu}$$

where μ = amplification factor, as before
V_g = grid potential relative to the cathode
V_s = segment potential relative to the cathode
Ψ_s = contact potential difference measured for the segment relative to the cathode.
We will use $\Psi_s = 2$ V since common values vary between 1.4 and 2.5 V for commercial tubes.
While μ is usually larger than 10 for commercial vacuum triodes, it is much smaller for VF tubes.

where t is the average grid transparency in the direction of the segment.

$a(x)$ corrects the segment line-of-sight distance $\sqrt{L^2 + X^2}$ to the value of Table A6.1 and is $[1 + (d_{fg} + d_{gs})/\mu d_{fg}]$. In what follows, $d_{gs} \approx d_{fg}$ and $a = 1 + 2/\mu$, with the result:

$$i_{seg} = \frac{ytAV^{3/2}}{al}\left(\tan^{-1}\frac{W_2}{L} - \tan^{-1}\frac{W_1}{L}\right)$$

This result is the current from a single filament; the total i_{seg} is calculated by summing the additional i_{seg} for each additional filament.

The grid current with all segments at zero voltage is found in the same way. The result for two symmetrical filaments is:

$$i_g(0) = I_{grid} = 2X\frac{2.335 \times 10^{-6}\ YV^{3/2}}{l}\left[\tan^{-1}\frac{W_1}{l} + \tan^{-1}\frac{W_2}{l} \times (1 - t)\right]$$

where I_{grid} = current to the grid
   $i_g(0)$ = current to the grid, segments off
   $2X$ = arises from 2 symmetrical filaments
   l = grid-to-filament distance
   Y = width of the grid in the direction of the filament and in the same units as w and d
W_1 and W_2 = distances to the edge of the grid from directly under one symmetrical filament
   V = filament-grid potential difference (the segments either are held at filament potential or charge down to filament potential)

The grid current $i_g(on)$ with segments on is:

$$i_g(on) = i_g(0)R + i_{seg}(1 - t)$$

where $1/R$ = ratio of the total effective grid area ($W_1 \rightarrow W_2$) to its area without the projected segment area, and
   i_{seg} = segment current for the on-segments.

Some calculated and measured values for digit 2 of a 7-segment VF tube with $\frac{1}{2}$-cm characters is given in Table A6.2 and is compared to measured values. Some of the error in the segment-current calculation error arises from using rectangular segment shapes; they actually have wedge-shaped ends. Also note that the optical grid transparency of 80% is measured for normally incident light whereas electrons are incident at an angle where the thickness is important. This may explain the electron transparency of only 76%.

Multiplication of space current by transparency to obtain segment current is only correct if $V_g = V_{seg}$.

Spangenberg[2] has shown that i_{seg} may be estimated more exactly as

$$i_{seg} = \frac{i}{1 + \delta^{-1}}\sqrt{V_g/V_s}$$

where δ is the current division factor.

This expression holds for grid transparency greater than 0.8 or 0.9.

Table A6.2.   **Calculated Segment and Grid Currents Compared to Experimental Values for Digit 2 of a Small VF Tube**

Parameter t	Calculated 87%	80%	76*	Experimental value, mA (+5%)
i_g(all)	1.45	2.23	2.68	2.2
$i_t = i_g$(all) $+ i_s$		3.78	4.15	3.5
i_a		0.216	0.205	0.160
i_d		0.216	0.205	0.180
i_g		0.162	0.154	0.122
i_b		0.274	0.260	0.241
i_c		0.274	0.260	0.252
i_e		0.206	0.195	0.170
i_f		0.206	0.195	0.155
i_s		1.55	1.474	1.282
$\mu = 2.5$				
i_g(off)	(used to calculate) $t = 76\%$ $t_{cpt} = 80\%$			2.6

The current division factor is almost equal to $t/(1 - t)$ and for t near 1 is $\simeq 1/(1 - t)$.

Then the above expression reduces to

$$i_{seg} = i_t [1 - (1 - t)] = i_t t$$

for $V_g = V_s$.

Note that static measurements of tubes designed for pulsed operation and nearly emission-limited operation such as the ISE single-filament FG94 can give misleading results. Significant filament cooling can occur at the digit under static test and this digit will become emission-limited. The grid will starve the segments and segment current will drop when the grid reaches its "natural" potential and beyond.

7

Liquid Crystal Displays

7.1 INTRODUCTION

The liquid crystal display (LCD) operates in a different fashion than the active display. It does not emit light of its own; instead, it must use an external source such as a radioactively excited phosphor, a VLED, a tungsten lamp, or the ambient light. Primary applications are low-power portable ones such as watches, calculators, or digital meters. Some other instruments also use LCDs.

7.2 LIQUID CRYSTAL PRINCIPLES

LCD Materials

The liquid crystal material is an organic compound. Ten percent of all organic materials form a phase called *mesophase* at some temperature between that required for liquid formation and that required for solid formation. Liquid crystals are mesomorphic. The resulting structure is birefringent (double-refracting) and turbid (scatters light). There are three mesophases of interest: smectic, nematic, and twisted nematic (or cholesteric). These are illustrated in Fig. 7.1. The twisted nematic is in widespread use. If the twisted nematic (tm) liquid is preferentially aligned on the front cell electrode (transparent) and then cross-aligned at 90° with respect to the top plate by the bottom electrode, the electric vector of polarized light will be rotated 90° in passing normally through the cell. Application of an electric field (ac or dc) destroys the rotation by tipping the molecules as in Fig. 7.2.

Long life requires ac operation since dc operation can lead to irreversible effects.

There are four classes of LC materials: azo-azoxy, Schiff bases, esters, and biphenyls. Some of the first watch displays used azoxy compounds and many now use Schiff base compounds. The biphenyls are of use in multiplexed applications. Table 7.1 shows some properties of these materials.

A less widely used technology is the use of dynamic scattering with nematic materials. No polarizers are used and the material becomes turbulent and scatters light when a field is applied. The appearance is a silver or colored background with gray numbers (it has had poor customer acceptance).

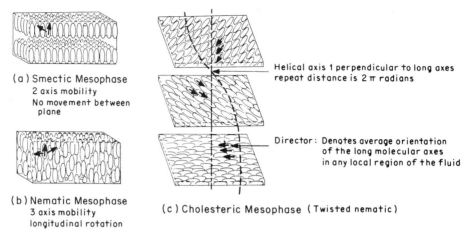

(a) Smectic Mesophase
2 axis mobility
No movement between
plane

Helical axis 1 perpendicular to long axes
repeat distance is 2π radians

Director: Denotes average orientation
of the long molecular axes
in any local region of the fluid

(b) Nematic Mesophase
3 axis mobility
longitudinal rotation

(c) Cholesteric Mesophase (Twisted nematic)

Fig. 7.1. Liquid crystal (LC) structure of the three LC mesophases. *(From Ref. 10.)*

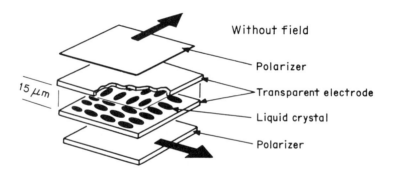

Without field

Polarizer

Transparent electrode

$15\,\mu m$

Liquid crystal

Polarizer

The applied field switches the 90° twist into a hemotropic structure.
The arrows represent the transmission axes of the polarizers.

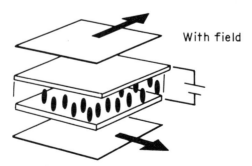

With field

Fig. 7.2. Schematic operation of a twisted nematic LC cell. *(From Ref. 10.)*

Table 7.1. Liquid Crystal Material Classes

Material	Notes
Azo-azoxy	DC operation risky, stable to H_2O
Schiff bases	Unstable and reactive
Esters	Nonreactive, hydrolyze, CO_2 evolves in operation
Biphenyls	Stable, relatively new, alignment problems

Optics

When the atoms that comprise a light source undergo dipolar transitions, they radiate randomly phased and polarized light. This situation is indicated schematically in Fig. 7.3a. Only the electric vector is considered, since only it produces polarization effects. If such light (also referred to as circularly polarized light) is incident on a polarizing material (polarizer), it is convenient to resolve it into parallel and perpendicular components (Fig. 7.3b). (Since the phase is random, n vectors average to $\sqrt{n}$ intensity.)

The polarizer severely attenuates the perpendicular component and transmits almost all the parallel component. When a second polarizer with its direction of

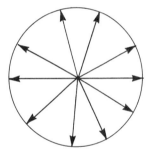

Fig. 7.3a. Schematic electric vector of circularly polarized light in propagation direction.

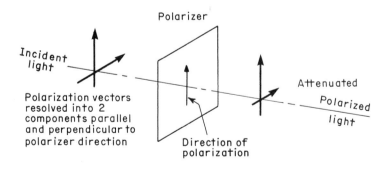

Fig. 7.3b. Polarization of circularly polarized light by means of a sheet polarizer (dichroic medium). Note that half the light is lost on polarization. The polarized half is attenuated by about 10% by the polarizer.

polarization at right angles is placed in the polarized beam, the beam will almost be blocked. The second polarizer is often referred to as an analyzer.

When a LC cell is placed between the polarizer and analyzer, an electrooptic light valve is created (Fig. 7.4).

When a point light source is used, a diffuser or collimator is used to obtain uniform illumination of the cell.

When reflected light operation is desired, the back of the polarizer may be marbled to provide the diffuser.

The brightness R(on) of a nonemissive cell is defined as the percentage contrast ratio between the light source with the LC cell and without the LC cell. For the case of a LC display used in the reflective mode, the cell brightness is often determined by using an MgO layer as the diffuser. The MgO returns 98% of the incident illumination.

The spectral brightness R_λ(on), or r(on), is the brightness for a particular spectral interval. The on- to off-contrast ratio is the reciprocal reduction in brightness by the off-state:

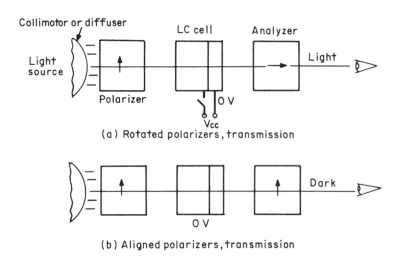

(a) Rotated polarizers, transmission

(b) Aligned polarizers, transmission

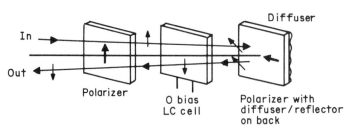

(c) Reflective – mode LC cell with polarizer arrangement shown. No bias (arrows) produces transmission and bias produces extinction.

Fig. 7.4. Liquid crystal polarizer arrangement.

$$C = \frac{R(\text{off})}{R(\text{on})}$$

The spectral contrast ratio C_λ is

$$C_\lambda = \frac{r(\text{on})}{r(\text{off})}.$$

For comparison purposes, a typical newspaper is found to have a 60% brightness and contrast ratio C of 5 and a magazine a brightness of 80% and a C of 15.[1]

A brightness measurement technique is shown in Fig. 7.5 where the source is a diffused area source and the detector makes cell in-out and on-off measurements. It is readily apparent that the source brightness is not a factor in the measurement of r or C.

An upper limit to the brightness and contrast ratio is determined from the polarizer-analyzer-diffuser combination alone. This contrast will be the highest achievable where on is the uncrossed and crossed is the off state. Introduction of the LC cell reduces these values. The polarizer order is important in the reflective case, when the two are different (Ref. 1). The display vendor will supply these details.

The contrast ratio is dependent on both voltage (Table 7.2) and viewing angle (Fig. 7.6). Figure 7.6 can be understood by considering the detector (or eye) of Fig. 7.5 moved at an angle Φ from the vertical and rotated in a circle around the display. Intersection of the $C = 3$ contour circle with the $\Phi = 30°$ figure are the 6 points where a contrast ratio of 3 is perceived.

Failure Modes

An incomplete list of failure modes includes:

- Bubbles in the liquid crystal
- Bubbles in the polarizer adhesive
- Increased leakage current
- Irreversible threshold voltage or threshold slope change

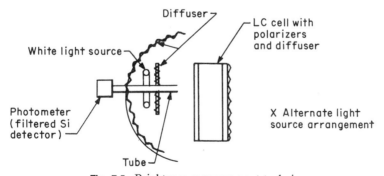

Fig. 7.5. Brightness measurement technique.

Table 7.2. On-Off Contrast Ratio
(Normal to Cell)

Reflective mode	
Voltage (rms)	Contrast ratio
3 V	4:1
4–6 V	20:1
Transmissive mode	
9–15 V	20:1

   An end-of-life criterion would be bubbles visible at ½ m or a 3 times increase in panel current (ac drive, dc test). Early wristwatch displays using an azoxy LC have been reported to have projected failures of less than 15% after 5 years (3% per year) under room temperature tests.[2] Improved units do as well at 40°C and also as well at 65°C storage.

7.3 LIQUID CRYSTAL DISPLAY

A 7-segment LC display is shown in Fig. 7.7. The construction in some respects is similar to the gas discharge display; tin oxide is used to form transparent electrodes on glass. The package is sealed with a screened glass frit using precision glass rod or spheres to obtain accurate and uniform spacing. Typical cell spacings are in the range of 10 to 12 μm (0.0005 in). An expansion joint closure after filling may be used in the form of a compliant epoxy. Alternatively, a flattened metal tube is sealed in

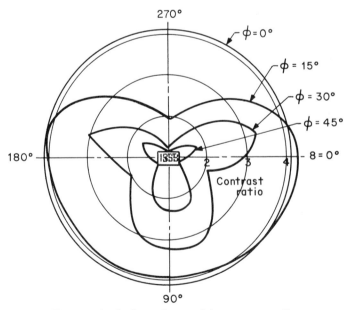

Fig. 7.6. Angle dependence of the contrast ratio.

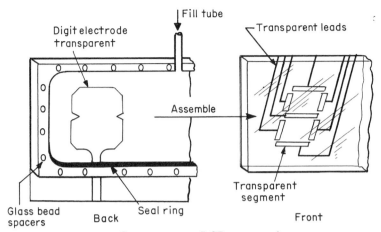

Fig. 7.7. Seven-segment LCD construction.

place with a glass frit. After cell filling, the tube is sealed by pinching off to form a cold weld.

Suitable polarizers are glued to the front and back to complete the assembly. Segment current levels are so small that contacts are not critical; pressure contacts are often used.

Operating Parameters

In addition to brightness and contrast ratios, response time, threshold voltage, and some derived figures of merit are used to specify LCDs. In particular, the viewing angle and temperature dependence of the above parameters are important.

The LCD is an ac-driven device; to aid parameter dependence and circuit considerations, it is necessary to consider the driving waveform properties.

An alternating rectangular drive pulse applied between a segment and digit address line will have an rms voltage V_{rms} equal to its peak value:

$$V_{\mathrm{rms}} = \frac{V_p}{\sqrt{2}} \quad \text{(see Sec. 7.4, under LC MUX Properties)}$$

If the segment and anode drive pulses of Fig. 7.8*a* is applied between a segment and anode, the resulting waveform is shown in Fig. 7.8*b*. As an example, if a 4-V (3 V worst case) battery supply is tripled (Fig. 7.9*a*), 12 V rms (9 V rms worst case) is available to drive the display. The double-pole, double-throw (DPDT) switch may be realized in Fig. 7.9*b* by alternating transistors *ad* on and *bc* on; *a*, *b*, or $3V_i$ is then varied to turn a cell on or off.

The LC threshold voltage V_{th} is defined by means of Fig. 7.10. The off-state voltage will be chosen to be less than V_{th}. The on-state voltage V_{on} will be selected to give a depolarization angle Φ less than 45°, which occurs about halfway up the curve. The ratio V_{on}/V_{th} will be selected to be between 2 and 4 by choice of LC material to give $\Phi < 30°$ in practice. Figure 7.10 also shows the voltage and viewing-angle dependence for a typical LC material.

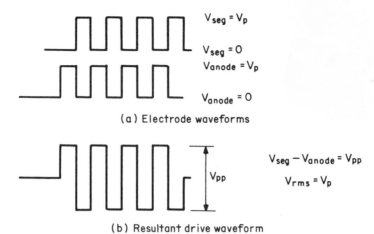

(a) Electrode waveforms

(b) Resultant drive waveform

Fig. 7.8. Driving waveform relationships.

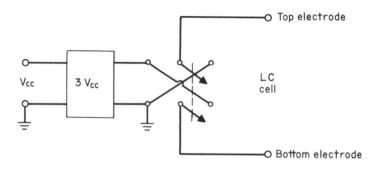

(a) Tripler driver

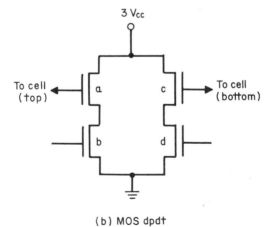

(b) MOS dpdt

Fig. 7.9. AC drive circuits.

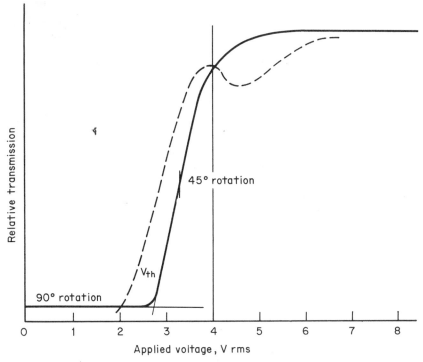

Fig. 7.10. Voltage dependence of tn cell transmission (11 polarizers). The solid curve is for normally incident light; the dotted curve is for light 30° off normal.

The rise times and fall times are defined to be those times for a 10 to 90% contrast ratio change at a specified angle. The decay time will generally exceed the rise time and can exhibit ''bounce'' where the cell is momentarily back on after switching off.

For a twisted nematic LC the threshold voltage is:[3]

$$V_{th} \propto d \sqrt{\frac{K_{22}}{\epsilon_0(\epsilon_{11} - \epsilon_1)}}$$

where d = cell spacing

K_{22} = LC parameter to be optimized

ϵ_{11} = dielectric constant along the LC molecule

ϵ_1 = dielectric constant across the molecule

It is seen that tn materials exhibit positive anisotrophy ($\epsilon_{11} - \epsilon_1 > 1$) and the cell spacing must be small and uniform.

Creagh[4] notes that LC mixtures allow threshold voltage tailoring; for example, 15% of a material known as PBAB mixed with another known as MBBA yields a 2 V rms threshold which is fully on at 2.8 V rms.

The rise and decay times are[5]:

$$t_{\text{rise}} = \frac{C_1 \eta d^2}{\epsilon V^2}$$

$$t_{\text{decay}} = \frac{C_2 \eta d^2}{K}$$

where C_1 and C_2 = constants
η = viscosity
d = spacing
K = LC parameter
V = rms voltage

Thus the d dependence is squared.

The temperature dependence T of the t's is partly in the viscosity η, where the η increases at low temperatures as:

$$\eta = \eta_0 \, e^{\Delta E / RT}$$

where η_0 = temperature-independent viscosity
R = 2.38 kcal/mole
ΔE = 4 to 10 kcal/mole

For example, Kahn and Burmeister[6] show how t_{off} varies from 450 ms at 0°C (η = 150 centipoises) to 55 ms at 50°C (η = 12 centipoises) for a certain biphenyl liquid crystal.

The temperature dependence of the threshold voltage has been developed by Kahn and Burmeister[6] in terms of threshold sharpness ρ, temperature coefficient δ, and voltage figure of merit M_v (Fig. 7.11).

The threshold sharpness ρ is related to the V_{on}/V_{th} ratio shown in Fig. 7.11 and is:

$$\rho = \frac{V_{90}}{V_{10}}$$

where V_{90} = rms voltage to give 10% of the maximum transmission
V_{10} = voltage for 90% of the maximum (9 : 1 contrast ratio)
ρ's of 1.4 are commonly attained for a vertical viewing angle.

The V_{th} decreases with increasing temperature, but ρ will be found to be maximum at some temperature which is specified as $\rho_{\max}(T)$. Figure 7.11 illustrates crossed polarizers viewed in transmission. The δ is defined for the increase in operating temperature to switch an off element (90% transmission) to on (10% transmission):

$$\delta_{\max}(T_{\text{LO}}, T_{\text{hi}}) = \frac{\rho(T_{\text{hi}}) - 1}{\rho(T_{\text{hi}}) + 1} \times \frac{200}{T_{\text{hi}} - T_{\text{LO}}}$$

as % per °C. The δ(0°C, 50°C) is in the range 0.35%/°C to 1.6%/°C for a variety of LC materials.[6]

Since a minimum δ does not necessarily mean a minimum ρ, a third figure of merit is M_v:

$$M_v = \frac{V_{90}(T_{\text{LO}})}{V_{10}(T_{\text{hi}})}.$$

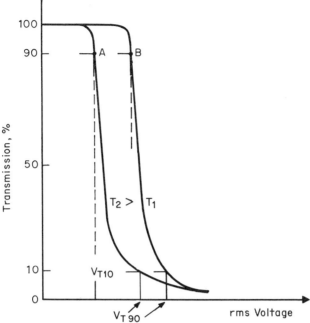

Fig. 7.11. Voltage dependence of transmission for LC with crossed polarizers illustrating definition of ρ, δ, and M_v, where $\rho = V_{T90}/V_{T10}$; $M_V = V_{T90}(T_1)/V_{T10}(T_2)$; and $\delta = [(A - B)/(\frac{1}{2})(A + B)] \times [100/(T_2 - T_1)]\%/°C$. *(After Ref. 6.)*

These parameters are also of importance for multiplexed displays where M_v will vary between 2 and 3 for typical LC materials.

Special Considerations

Some precautions in using LC devices are:

1. Protect against freezing.
2. Avoid boiling and storage above 100°C.
3. Double images will result if a mirror reflector is used with dark numerals on a transparent background.
4. Temperature-controlled thin-film heaters are commercially available to stabilize displays at low temperatures. Their use can double the display cost.
5. Colored front polarizer combinations can be used. (Some manufacturers may be able to supply these.)
6. Colors should be user tested prior to introduction.

7.4 MULTIPLEXING PRINCIPLES

Multiplexing design consists of three parts: determination of the optimum pulse conditions; designing the pulse sets to meet these conditions; and designing the circuits to generate the pulse sets. Approach to optimum pulse conditions will be demonstrated here with some pulse sets from the literature. The general cases are left to Appendixes 7.1 and 7.2. Circuit designs must be obtained from the LC supplier.

LC MUX Properties

There will be a frequency range over which the tn or dynamic scattering materials will respond to the rms voltage rather than the instantaneous. For example, the contrast decay time for the tn material is of the order of 100 ms, and frame times T will be about 50 ms to avoid flicker.

Where N lines are strobed, the duty cycle will be $1/N$ and the dwell time will be T/N.

The rms voltage $\hat{V}$ is the square root of:

$$\hat{V}^2 = \frac{\int_0^T V_T^2 \, dT}{T}$$

As an example, Fig. 7.12 shows a rectangular pulse $T/2$ long by V_0 high. The rms voltage is

$$\hat{V} = \frac{V_0}{\sqrt{2}}$$

If an attempt is made to strobe the 3-digit, 2-segment display of Fig. 7.13a with the pulses of Fig. 7.13b, it is seen that all rms voltages are the same whether a full-select, half-select, or nonselect condition occurs. (See Fig. 7.13c for definition.)

A 3-digit by 3-segment display (Fig. 7.14a) is considered after the fashion of Ref. 7. The pulses are shown in Fig. 7.14b. If a single-segment select pulse is used to select a_1 which is $\sqrt{3}$ times the digit select pulse, the $\hat{V}_{on}/\hat{V}_{off} \geq \rho$ is found to be 1.82. The $\hat{V}_{on}/\hat{V}_{off}$ ratio is also dependent on the segment number since it is 1.47 when two segments are selected.

It is also select-pulse amplitude dependent; for $V_a = 2V_D$, $\hat{V}_{on}/\hat{V}_{off}$ for two selected segments is 1.91. There clearly is an optimum select-pulse height.

Figure 7.15 shows an idealized transmission-threshold curve where V_{th} is normal-

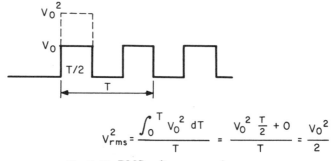

$$V_{rms}^2 = \frac{\int_0^T V_0^2 \, dT}{T} = \frac{V_0^2 \frac{T}{2} + 0}{T} = \frac{V_0^2}{2}$$

Fig. 7.12. RMS voltage example.

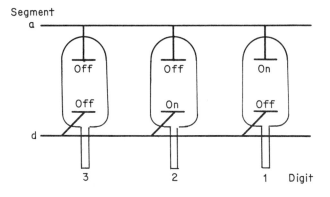

(a) 3-digit, 2-segment LC display

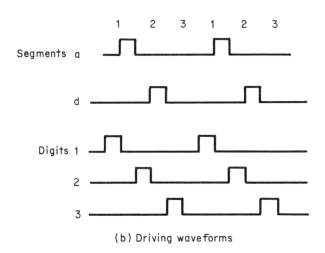

(b) Driving waveforms

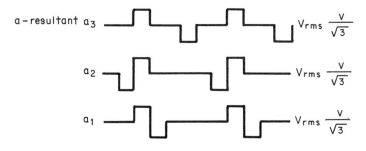

(c) Resultant a-cell waveforms

Fig. 7.13. RMS voltages developed by standard multiplexing scheme showing no-selection.

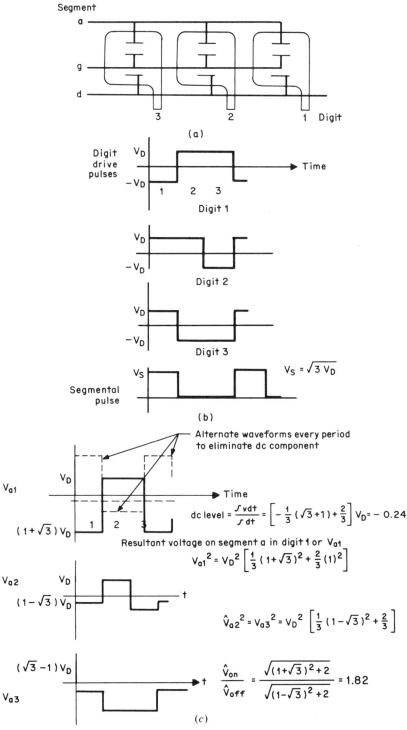

Fig. 7.14. (*a*) 3-digit, 3-segment display example. (*b*) Drive pulses. (*c*) Resultant waveforms.

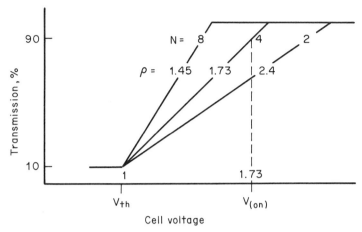

Fig. 7.15. Idealized threshold curves (parallel polarizers). *(After Ref. 7.)*

ized to 1 V. Then $\rho = \hat{V}_{on}/\hat{V}_{th} = \hat{V}_{on}$. On the curve is shown the maximum N for a given ρ based on the theory explained in Appendix 7.1.

Note that the new digit entry times will exceed the rise time. In fact, it will be at least one rise time plus two frame times on the average and probably nearer three.

Multiplexed digital power requirements usually must be measured directly to account for all circuit capacitances. Estimates may be made by assuming that all the reactive power is lost.

REFERENCES

1. T. J. Scheffer and J. Nebring, *Proc. of the Soc. for Inf. Disp.*, **18** (1977), p. 33.
2. Seiko Watch Co.
3. E. Guyon and W. Urbach, "Anchoring Properties and Alignment of Liquid Crystals," in A. R. Kmetz and F. K. von Willisen (eds.), *Non-Emissive Electro-optic Displays*, Plenum, New York (1975), p. 261.
4. L. T. Creagh, *Proc. IEEE*, **61** (1973), p. 814.
5. L. M. Blinov, *Usp. Fiz. Nauk*, **114** (1974), p. 671.
6. F. J. Kahn and R. A. Burmeister, Jr., "Temperature Dependence of Multiplexed Twisted Nematic Liquid Crystal Displays," in A. R. Kmetz and F. J. von Willisen (eds.), *Non-Emissive Electro-optic Displays*, Plenum, New York (1975), p. 261.
7. A. R. Kmetz, "Matrix Addressing of Non-Emissive Displays," in A. R. Kmetz and F. K. von Willisen (eds.), *Non-Emissive Electro-optic Displays*, in Plenum, New York (1975), p. 261.
8. P. M. Alt and P. Pleshko, *IEEE Trans. Electron Devices*, **21** (1974), p. 146.
9. H. Kawakami, Y. Nagae, and E. Kaneko, *Conf. Record 1976 Biennial Display Conference, IEEE*, New York (1976), p. 50.
10. Shelley, Associates.
11. G. G. Barma, *Conference Record of 1976 Biennial Display Conference*, Society for Information Display, Los Angeles (1976), p. 29.

APPENDIX 7.1
An Optimum Set of Multiplexing Voltages[7,8]

Figure A7.1 shows one of several rows and N columns; at any time the row has applied either V_1, a nonselect, or V_2, a select voltage, referenced to the column nonselect voltage V_4. The column select voltage is V_3. These voltages are to be determined. In this example, n selected cells and $N - n$ nonselected cells are chosen. If a nonselected cell has a column-select pulse applied to select a cell on another row (such as row N in the figure), it is known as a half-selected cell. Three voltages are now considered: (1) the root mean square (rms) voltage $\hat{V}_s$ across a selected cell such as the cell of column 1 (the caret represents the rms value); (2) the voltage $\hat{V}_{ns}$ across a nonselected cell such as on column 2; (3) the voltage $\hat{V}_{hs}$ across a half-selected cell:

$$\hat{V}_s^2 = (V_2 + V_3)\frac{1}{N} + \frac{n-1}{N}(V_2 - V_4)^2 + \frac{N-n}{N}(V_1 - V_4)^2$$

$$\hat{V}_{ns}^2 = \frac{n}{N}(V_2 - V_4)^2 + \frac{N-n}{N}(V_1 - V_4)^2$$

$$\hat{V}_{hs}^2 = (V_1 - V_3)^2\frac{1}{N} + \frac{n}{N}(V_2 - V_4)^2 + \frac{N-n-1}{N}(V_1 - V_4)^2$$

To eliminate the n-dependence, $\hat{V}_s^2$ and the worst case off-state (V_{hs}^2) are rearranged:

$$N\hat{V}_s^2 = (V_2 + V_3)^2 - (V_2 - V_4)^2 + N(V_1 - V_4)^2 + n((V_2 - V_4)^2 - (V_1 - V_4)^2]$$
$$N\hat{V}_{hs}^2 = (V_1 - V_3)^2 - (V_1 - V_4)^2 + N(V_1 - V_4)^2 + n[(V_2 - V_4)^2 - (V_1 - V_4)^2]$$

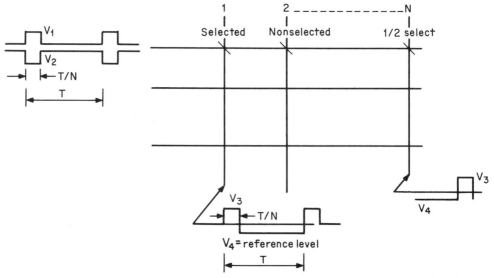

Fig. A7.1. Row column matrix showing selected, nonselect, and half-select cells. Also shown are the definitions of V_1, V_2, V_3, and V_4.

Fixing the reference level $V_4 = 0$, the condition $V_2 = -V_1$ yields n-independence and on-state ($\hat{V}_s$) and off-state ($\hat{V}_{ns}$) differences.

Substituting $V_1 = -V_2$, and adding and subtracting $2V_2V_3$ to V_3^2, the following ratio is obtained:

$$\frac{\hat{V}_{on}^2}{\hat{V}_{off}^2} = 1 - \frac{4V_2V_3}{2V_2V_3 + V_3^2 + NV_1^2}$$

Dividing the numerator and denominator of the fraction by V_3^2, the result is

$$\frac{\hat{V}_{on}^2}{\hat{V}_{off}^2} = 1 - \frac{4\,V_2/V_3}{1 + 2V_2/V_3 + N(V_2/V_3)^2}$$

The ratio is maximized by differentiating with respect to V_1/V_3 and setting the derivative to zero. A result is

$$V_3 = \sqrt{NV_2}$$

and substituting this result [and recalling $1 - N = (1 - \sqrt{N})(1 + \sqrt{N})$] in the original expressions gives

$$\frac{\hat{V}_{on}}{\hat{V}_{off}} = \sqrt{1 + \frac{2}{\sqrt{N} - 1}} = \sqrt{\frac{\sqrt{N} + 1}{\sqrt{N} - 1}}$$

This function is the one plotted on the threshold curves of Fig. 7.14, where $\hat{V}_{on}/\hat{V}_{off} > V_{on}/V_{th}$.

The voltage V_3 is called the address pulse or select pulse and V_2 is called the information pulse or data pulse.

APPENDIX 7.2
Pulse Design

Kawakami, Nagae, and Kaneko[9] show how to derive pulse sets from optimum pulse-voltage conditions. Their pulse designs are also relatively straightforward to implement in MOS form. They show that the results derived in Appendix 7.1 can be generalized to symmetrical cell waveforms. To follow their deviation, let $V_{21} = -V_3$ (where V_3 is defined in Appendix 7.1), $V_{22} = -V_4 \neq 0$, $V_{11} = -V_2$, $V_{12} = -V_1$. Then at one T/N interval:

$$V_s = V_{21} - V_{11}$$
$$V_{hs} = V_{22} - V_{11}$$
$$V_{ns} = V_{22} - V_{12}$$

and a selected row on a nonselected column is $V_{hs} = V_{21} - V_{12}$.

They derive the result for $V_4 \neq 0$ ($V_{12} \neq 0$):

$$V_{12} = V_{11} \pm \frac{2}{b}V_0 \qquad b = 1 + \sqrt{N}$$

$$V_{21} = V_{11} \pm V_0 \qquad\qquad\qquad\qquad (A7.1)$$

$$V_{22} = V_{11} \pm \frac{1}{b}V_0$$

(*Note:* If $V_{22} = 0$ then $V_{11} = -\frac{1}{b}V_0$, and $V_{12} = -V_{11}$, $V_{21} = -\sqrt{N}\ V_{11}$ as before).

Then the result for time T_1/N is:

$$V_s(T_1) = V_{21} - V_{11} = \pm V_0$$

$$V_{hs} = V_{21} - V_{12} = \pm\left(1 - \frac{2}{b}\right)V_0$$

$$V_{22} - V_{11} = \pm\frac{1}{b}V_0$$

$$V_{ns} = V_{22} - V_{12} = \pm\frac{1}{b}V_0$$

They show, as in Fig. A7.2, a possible set of pulses for $V_4 = V_{22} = 0$. A state diagram for the case of Fig. A7.2.1 is shown in Fig. A7.3.[2]

They then show that the largest applied voltage V_0 can be reduced to ½, by use of the state diagrams of Fig. A7.4 to determine the driving waveform set. It is seen that these waveforms may be readily selected by use of a DPDT switch arrangement which has one full and two intermediate states.

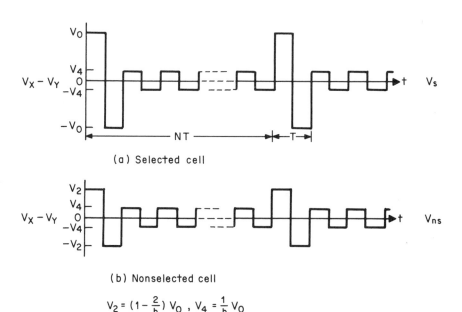

(a) Selected cell

(b) Nonselected cell

$$V_2 = \left(1 - \frac{2}{b}\right)V_0\ ,\ V_4 = \frac{1}{b}V_0$$

Fig. A7.2. Waveforms across a selected and a nonselected cell.

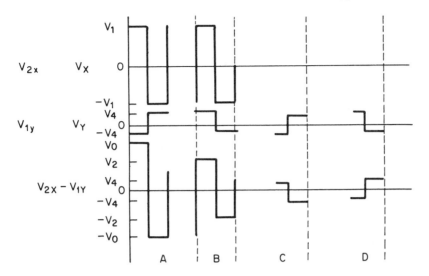

A: fully selected state
B: half selected state on a selected row
C: half selected state on a selected column
D: nonselected state

$$V_1 = (1 - \frac{1}{b}) V_0, \quad V_2 = (1 - \frac{2}{b}) V_0, \quad V_4 = \frac{1}{b} V_0$$

Fig. A7.3. State diagram of waveforms, where V_{11} in Eq. (A7.1) is equal to $-1/b(V_0)$ and V_{11} in Eq. (A7.1) is equal to $1/b(V_0)$. *(Ref. 9.)*

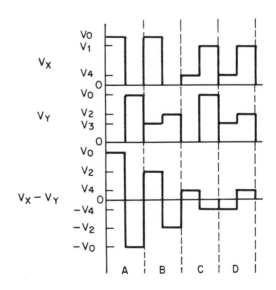

Figure A7.4. State diagram of waveforms, where V_{11} in Eq. (A7.1) is equal to V_0 and V_{11} in Eq. (A7.1) is equal to 0. *(Ref. 9.)*

$$V_1 = (1 - \frac{1}{b}) V_0, \quad V_2 = (1 - \frac{2}{b}) V_0, \quad V_3 = \frac{2}{b} V_0, \quad V_4 = \frac{1}{b} V_0$$

8

The Incandescent Display and the Cathode-Ray-Tube Display

8.1 THE INCANDESCENT DISPLAY

Introduction

The incandescent display operates like an ordinary light bulb; the display segments are filaments heated to incandescence. The display differs from an incandescent light in that it operates at a lower temperature. The advantages of the incandescent display are high brightness at reasonable power and voltage. It can also be filtered to provide a full range of colors. It has the disadvantage of requiring increased voltages for multiplexed operation at the same high brightness levels. Applications include displays in cash register, scales, gasoline pumps, and some instruments.

Operation

An incandescent metal will radiate a blackbody (BB) spectrum (Fig. 8.1) which depends on its temperature and emissivity. Note from the figure that all colors are present in the 1400 K spectrum and that any color may be selected by filtering (with a corresponding loss in brightness). The BB spectrum has much of its radiation below eye sensitivity. As a result, the unfiltered display segment will appear yellow rather than white.

The turn-on and turn-off times will be long compared to the displays considered earlier. The time to 80% brightness is quoted by one manufacturer as 20 ms.[1] Nearly all the input power must be radiated as thermal radiation for a well-designed device, and is

$$\frac{dQ}{dt} = A\epsilon\sigma(T^4 - T_3^4)$$

where $\dfrac{dQ}{dt}$ = rate of heat gain or loss in watts

A = effective radiating area

ϵ = filament emissivity

T = filament temperature

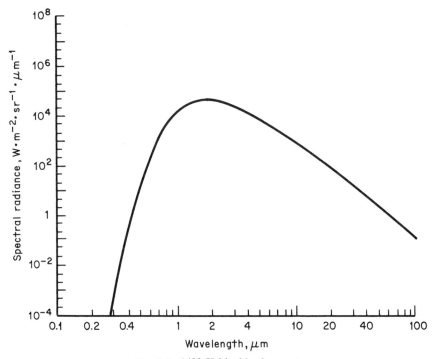

Fig. 8.1. 1400-K blackbody spectrum.

T_3 = temperature of the surroundings

σ = Stefan-Boltzmann constant (5.6697×10^{-8} W·m^{-2}·K^{-4})

From the definition of the specific heat, C_p: $dQ = C_p\, dT$. The equation for the radiation gain or loss for the hot filament is

$$C_p \frac{dT}{dt} = A\,\epsilon\sigma(T_0^4 - T_3^4)$$

The filament behavior according to this equation is described in Appendix 8.1.

When a cold filament is first turned on, a large current spike occurs. This spike is equal to the driving voltage divided by the cold resistance (Appendix 8.1). Either the switching transistor must be designed to accommodate this transient or a poorly regulated power supply must be used.

Segment luminance and current as a function of voltage are given on some manufacturers' data sheets (Fig. 8.2). Calculation methods are described in Appendix 8.1.

Construction

The display tube is constructed with stainless steel or nickel support wires in a vacuum tube or flat package. The coiled filaments are spot-welded to the supports; a blackened nickel or insulating background can also be provided for contrast enhancement. The tube is generally under a vacuum and a getter is provided to ensure long life.

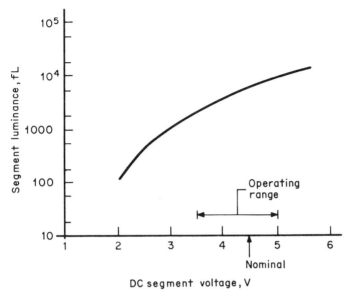

Fig. 8.2. Segment luminance as a function of segment voltage. *(Adapted from Ref. 2.)*

Reliability and Environmental Considerations

Failure modes for this display have not been described. Filament evaporation is negligible for such low-temperature operation. Vacuum life with a getter should exceed two years and may easily reach five years. Since many manufacturers recommend a broken filament circuit (if the current is zero on a selected filament, the digit is blanked), one possible failure mechanism is broken filaments.

Operating temperature ranges are typically −55 to +100°C.

The displays are rugged; some manufacturers specify performance under reasonably high environmental stresses per MIL-STD-202 test methods for fresh displays.[1]

Failure rates of 0.7 to 0.8% per 1000 hours (95% confidence when tested in a particular circuit and 4X accelerated test mode) are quoted by one manufacturer.[2]

Tungsten filament life is normally derated or rerated by a twelfth-power law. For example, if a 4.5-V filament is operated at 5 V, its life is shortened by $(5/4.5)^{12}$, or 3.5, times.

Multiplex Operation

The incandescent filament display may be multiplexed. Segment isolation is accomplished through the use of an external series diode in each segment line. Significant filament heating and cooling can occur unless high multiplex rates are used. Less than 80% cooling will occur within .02 of the time to heat to 80% brightness. If the latter time is 20 ms, the segment off time should be less than 0.4 ms. For an $\frac{1}{8}$ duty cycle a safe digit repetition rate would be 1/.04 ms, or 25 kHz. One manufacturer[1] recommends a 5-kHz rate for 10 multiplexed digits.

The rms power must be maintained the same as the dc power requirement to maintain the same brightness. If the duty cycle is $1/N$, then $V_{rms} = V_{dc}\sqrt{N}$ and $i_{peak} = Ni_{dc}$. Note that the transient peak for cold starting will exceed i_{peak} by about 4 times.

8.2 THE CATHODE-RAY-TUBE DISPLAY

Introduction

When the number of characters exceeds 2000, the CRT display is unsurpassed. It finds its greatest application in computer interface applications.

As the classical display, much of the design technique is embodied in textbook descriptions and manufacturers' subsystems. Yoke manufacturers supply deflection know-how; transformer manufacturers supply the arcana necessary to resonate their flyback transformers and so on for various subsystems throughout the display system.[3] Here will be noted some phosphor choices, CRT figures of merit, and some circuit preliminaries.

Cathode-Ray-Tube Considerations

A number of factors must be considered to design a CRT display. Some are shown in Fig. 8.3. It is noted that environmental stress is generally not a factor since the display usually is used in an air-conditioned room.

The phosphor is chosen to be white or green. Table 8.1 shows some choices available from the JEDEC listing. The cathode-ray tube (CRT) can generally be specified to have any phosphor desired. Other CRT factors are shown in Table 8.2. The designer should work closely with the CRT manufacturer to optimize these factors in his application.

Notes on Character Display

Figure 8.4 is an example of how the screen is divided to allow a series of lines of characters to be displayed. Note that 20 to 25% is off the screen at the top and also

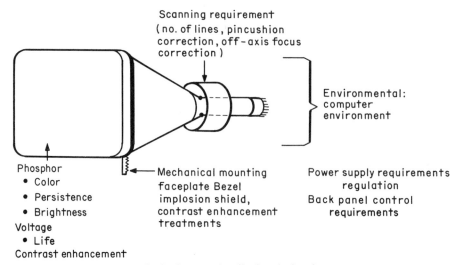

Fig. 8.3. Cathode-ray-tube display design factors.

Table 8.1. Phosphor Characteristics
(Standard JEDEC Registered Phosphors)

EIA phosphor	Emission color fluorescence/ phosphorescence	Composition	Application
P1	Yellowish-green	Zinc silicate:manganese	Used in cathode-ray oscillographs and radar
P4	White	Zinc sulfide:silver and zinc cadmium sulfide:silver	Used in monochrome TV picture tube
P31	Green	Zinc sulfide:copper	Used in cathode-ray oscillographs
P32	Purple-blue/yellow-green	Calcium magnesium silicate:titanium zinc cadmium sulfide:copper	Used for radar display
P33	Orange	Magnesium fluoride:manganese	Used for radar display
P34	Bluish-green/yellow-green	Zinc sulfide:lead:copper	Radar
P35	Yellowish-green/ greenish-blue	Zinc sulfide selenide:silver	Oscillograph
P36	Yellowish-green	Zinc cadmium sulfide:silver:nickel	Flying spot scanner
P37	Greenish-blue	Zinc sulfide:silver:nickel	Flying spot scanner
P38	Orange	Zinc magnesium fluoride:manganese	Radar display
P39	Yellowish-green	Zinc silicate:manganese:arsenic	Low-repetition displays
P40	White/yellowish-green	Zinc sulfide:silver zinc cadmium sulfide:copper	Low-repetition displays
P41	Orange-yellow	Zinc magnesium fluoride:manganese calcium magnesium silicate:cerium	Visual displays
P42	Yellowish-green	Zinc sulfide:copper, zinc silicate:manganese:arsenic	Low-repetition, high-brightness display
P43	Yellowish-green	Gadolinium oxysulfide:terbium	Visual displays
P44	Yellowish-green	Lanthanum oxysulfide:terbium	Visual displays
P45	White	Yttrium gadolinium oxysulfide:terbium	For applications requiring high brightness
P46	Yellowish-green	Yttrium aluminate:cerium	High efficiency for scanning and recording
P48	Yellowish-green	70:30 mix P46:P47	Flying spot scanners
P49	Green, red		Graphics and alphanumeric displays

off at the side where the retrace occurs. It is convenient to locate the black reference pedestal and sync pulses here also. If a 5 × 7 PROM (programmable read-only memory) is used as a character generator, a 6 × 7 character block can be used to allow line and character separation.

Figure 8.5 shows a block diagram of how a master clock is used to generate master sync and to load line-at-a-time character information from the keyboard or computer into a page-storage random-access memory (RAM). This procedure allows continuous updating at the frame rate. The master clock or dot frequency is clearly the total lines times the line frequency times the cells per line times the dots per cell.

Table 8.2. Figures of Merit

Size:
    Catalogue list and availability
    Mounting space
    Amount of data to be displayed and
    Viewing distance (font size)
    Rule of thumb
        3 mm character at 50 cm
        6 mm character at 1 m
Contrast:
    Whole-screen contrast
     factor: reflected ambient lighting (Use light-absorbing filter)
    Small adjacent area contrast
     factor: focus character of CRT
Resolution:
    Provide sufficient contrast between details
        If beam current increases, r_{spot} increases; requires special attention to deflection coil selection
        Larger-diameter gun reduces design problems at expense of higher power.
Luminance of display (in addition to adequate contrast):
    resolution-brightness tradeoff
        Best phosphor is most efficient.
        High second anode potential is desirable.
Flicker:
    Greater than 60-H_z repetition rate
    Choose persistence to reduce flicker fatigue but allow adequate erasing (avoid lag).
    Judge flicker (under actual conditions of use):
        Watch for annoying beat frequencies between fluorescent light and display.
        Use large number of observers.

SOURCE: Ref. 5.

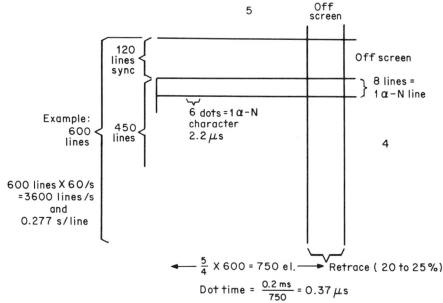

Fig. 8.4. Screen partition of α-N display.

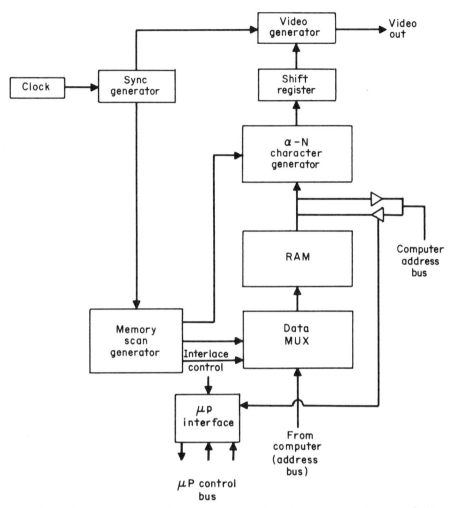

Fig. 8.5. Block diagram of alphanumeric video character generator. *(After Ref. 4.)*

In the figure, the dot counter divides the clock frequency by dots per cell and steps the cell counter and so on.

REFERENCES

1. Refac Electronics Corporation.
2. Radio Corporation of America.
3. H. H. Poole, *Fundamentals of Display Systems,* Spartan Books, Washington, D.C. (1966).
4. B. Matic and L. Trotlier, *Electron. Des.,* **19** (1977) p. 68.
5. Clinton Electronics Corporation.

APPENDIX 8.1
Incandescent Display Parameter Estimation

A8.1   FILAMENT TEMPERATURE VARIATION WITH VOLTAGE

A convenient means of estimating the variation of temperature with voltage (to allow estimates of brightness variation, for example) is to note the temperature varies as the $\frac{1}{4}$ power of the filament power:

$$T_o = P^{1/4} = \left[\frac{V_o^2}{R_3}\left(\frac{T_o}{T_{A3}}\right)\right]^{1/4}$$

where V_o = operating voltage
$\quad R_3$ = room temperature resistance
$\quad T_o/T_A$ = variation of resistance with temperature for non-sag tungsten wire
$\quad T_o$ = operating temperature
$\quad T$ = room temperature

Then if the voltage is increased from V_{o1} to V_{o2}, the resulting equation is obtained:

$$\frac{T_{o1}}{T_{o2}} = \frac{V_{o1}^{1/2}\, T_{o2}^{1/4}}{V_{o2}^{1/2}\, T_{o1}^{1/4}}$$

and

$$\frac{T_{o1}}{T_{o2}} = \left(\frac{V_{o1}}{V_{o2}}\right)^4$$

A8.2   SEGMENT CURRENT-VOLTAGE RELATION

Since the segment resistance is temperature-dependent, the current i at a given voltage V will vary as

$$i(T) = \frac{V}{R(T_o)}$$

where $R(T_o)$ varies as $R_3(T_o/T_3)$, as in Sec. A8.1.

The current at the nominal voltage will determine $R(T_o)$ and can be taken from the data sheet or direct measurement. If the filament is not non-sag tungsten, slight deviations from linearity will be experienced.

Then from Sec. A8.1:

$$\frac{i_1(T_{o1})}{i_2(T_{o2})} = \frac{V_{o1}\, R_2(T_{o2})}{V_{o2}\, R_1(T_{o1})} = \frac{V_{o1}}{V_{o2}}\frac{T_{o2}}{T_{o1}}$$

Then if the current is changed from i at V_{o1} by changing the voltage to V_{o2} the new current i_2 is:

$$i_2(T_{o2}) = \left(\frac{V_{o2}}{V_{o1}}\right)^{0.6} i_1(T_{o1})$$

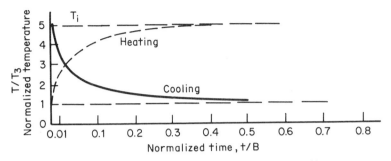

Fig. A8.1. Heating and cooling of an incandescent filament.

A8.3   SEGMENT HEATING AND COOLING TIMES

The equation for segment heating and cooling given in Sec. A8.2 is solved by partial fractions and integration as:

$$t = \frac{C_p}{A\epsilon\sigma T^3}\left[-\tfrac{1}{2}\tan^{-1}\left(\frac{T}{T_3}\right) + \tfrac{1}{2}\ln\left(\frac{T/T_3 - 1}{T/T_3 + 1}\right)\right] + C$$

with the boundary condition for cooldown:

$$\frac{T}{T_3} = \frac{T_o}{T_3} \qquad \text{at } t = 0$$

Then the solution for cooldown is:

$$t = \frac{C_p}{2A\epsilon\sigma T_3^3}\left[-\tan^{-1}T/T_3 + \tfrac{1}{2}\ln\left(\frac{T/T_3 - 1}{T/T_3 + 1}\right) - \tfrac{1}{2}\ln\left(\frac{T_o/T_3 - 1}{T_o/T_3 + 1}\right)\right]$$

This result is plotted as a function of T/T_3 in Fig. A8.1 and normalized to t/B. B is just $C_p/2A\epsilon\sigma T_3^3$ and $A\epsilon$ may be calculated from $W = \epsilon A\sigma T^4$.

For example, if $W = 0.1$ watt (5 V × 20 mA), $A\epsilon$ is found to be 2.6×10^{-3} cm^2 (2.6×10^{-7} m^2), for $T_o/T_3 = 5$ ($\cdot B$ is about 40 ms in this case). The heat-up time (dotted curve) is found by subtracting the value less 1 of T/T_3 at t from 5 (or T_o/T_3).

A8.4   COLD STARTING CURRENT

When the filament is cold, the inrush current will be:

$$i = \frac{V}{R_3} \qquad \text{and} \qquad i(t) = \frac{V}{R_3}(t)$$

where R_3 is found from R_o by using the known operating temperature. As an example, if $T_o = 1500$ K, 80% brightness is reached in 18 ms ($t/B = .27$), V_{dc} is 4.5 V, but V_{mux} is 13 V and $R_3 = 50$ Ω. Then, $i(t) = {}^{13}\!/_{50}\,(T/T_3)$.

The curve of Fig. A8.2 may be constructed from Fig. A8.1.

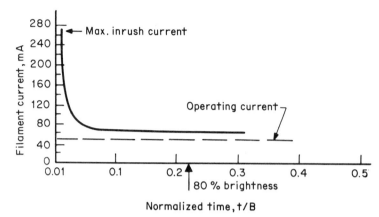

Fig. A8.2. Cold starting current for a tungsten filament.

INDEX

Abnormal glow region, 60
Absorption:
  light, 106, 108
  optical, 104
Accelerated life test, 20
Acceptance, lot, 20
Accuracy, and legibility, 50
Acuity, visual, 41
Acutance (sharpness), 51
Addressing, methods of, 14
Adhesive for mounting, 18
Aging (*see* Burn-in, VF tube aging)
Aligned liquid crystal, 137
Allocation, display fraction of, 13
Alloys, GaAsP, 99–110
Alphanumeric display, 116
  CRT block diagram of, 163
  VLED, 110, 114
  VLED drive, 120
Amplification, vibration, 22
Analyzer, 139–155
Angle:
  critical, 42
  depolarization, 143–145
Anode:
  common, VLED, 116
  gas discharge, 59
  indium oxide, 72
  tin oxide, 72
  transparent, 72
  in VF construction, 124
Anode drive pulse, LCD, 143
Anode drivers, gas discharge, 79
Antimoisture agents, 129
Antireflective coating, 43, 106
Aperture, standard candle, 33
Apparent brightness, 33
  MUXed, 16
Applications, display:
  business, 3, 4
  consumer, 3, 4

Applications, display (*Cont.*):
  industrial, 4
  medical, 4
  military, 4
Argon, 65
Aspect ratio, 5, 51
  character, 5
  segment, 5
Azo-azoxy, LCD, 137, 139

Ba, Sr, Ca, O (reduced triple carbonate), 129
Background, 29
  black or gray, 43
  illumination of, 33
  reflected, 30
Ballast resistor, 60
  elimination of, 68
Band gap:
  indirect, 101
  pseudodirect, 101
Bathtub failure curve, 75
Bezel, 18, 42
  recessed, 43
Biphenyls, 137, 139, 146
Birefringent structure, 137
Blackbody (BB) spectra, 33, 34
  1400 K, 157, 158
  2042 K, 33
    Planck's equation for, 33, 34
Blemishes, 23
  cost-effective, 13
  specifiable, 13
Bond failure, elimination of, 114
Bounce, LCD, 145
Breakdown, gas, 61
Breakdown voltage, gas discharge, 66
Brightness, 29
  apparent, 33
    MUXed, 16
  current density, scaling, 9

Brightness (*Cont.*):
  LCD, 141
  of nonemissive cell, 140
  rule-of-thumb, 13
  spectral, 140
  subjective, definition of, 57
  VF uniformity and, 125
  VLED, 108–110
Brightness degradation, minimized, 114
Brillouin zone of GaP, 100
Build-up (of luminance), definition of, 55
Burn-in:
  drive circuit aided, 74
  mercury coating in, 74
  VF aging and, 129
  VLED, 113, 114
Burn-in rack, 20
  testing on, 22

Calculator circuit, gas discharge display,
  82
Carrier concentration, optimum, 105
Cathode:
  common, VLED, 116
  film, 73
  gas discharge, 59
Cathode current, definition of, 64
Cathode glow, 59
Cathode off-state voltage, definition of, 64
Cathode-ray tube (CRT), 160
  page-storage memory, 161
  screen, division of, 160, 162
Cathode-ray-tube character generator, 1
  block diagram, 163
Cathode-ray-tube display, 160–163
Cathode-ray-tube figures of merit, 160, 162
Cathode-ray-tube phosphor, 160–163
  choice of, 160
Cathode supply voltage, definition of, 64
Cathodoluminescence, 104
Centroid, use of, in VLED spectrum, 108
CFF (*see* Critical fusion frequency)
Characters:
  size of, 51
  slanted, 5
  stacking of, 7
Charge carrier fate chart, 102
Cholesteric mesophase, 137
Circuit, calculator, gas discharge display,
  82
Circular polarizer, 43
Clamping, 27
Clock, multiphase, 81

Clock display, 7, 9
Coating:
  antireflection, 106
  compliant, 114
  conformal, 106
Cold weld, 143
Collimator, 140
Color:
  definition of, 55
  glow discharge, 68, 70
  VLED, 108
Color boundary, 44
Color contrast, 44, 45
Color names, 44, 45
Color temperature, 30
Column-select pulse, 152–155
Comfort level factor, 50
Common anode, VLED, 116
Common cathode, VLED, 116
Comparison, display:
  CRT and gas discharge dot matrix
    tube, 16
  gas discharge, VLED, and VF, 19
  LCD, ECD, and EPID, 17
  between technologies, 14, 15
  VLED and gas discharge, 18
Comparison methods, 13
Compliant coating, 114
Conformal coating, 106
Confusion matrix, 51
Connectors, 18
  display, 22
Construction:
  incandescent display, 158, 159
  LCD, 142
Construction techniques:
  gas discharge display, 70, 72
  VF display, 127–129
Contact potential difference (CPD),
  126
Contacts:
  hard, 74, 129
  pressure, 143
Contrast:
  bar-to-bar, 30, 31
  color, 44, 45
  definitions of, 30, 32
  display, 31
  minimum, 48
  for numerics, 50
  positive or negative, 29
  symbol-background color, 49
Contrast enhancement, 18, 29, 39–51,
  158

Contrast ratio:
  LCD, 140
  luminance, definition of, 56
  spectral, LCD, 141
Contrast reduction, 42
Conversion, cd to fl, 57
Conversion factors, luminance, 40
Cosmetic appearance, gas discharge, 75
Cosmetic specification, 75
Cosmetic standards, 23
Cover:
  reflection by, 43
  shuttered, 43
  VF, 129
Critical angle, 42
Critical fusion frequency (CFF), 16, 53, 55
Cross-aligned liquid crystal, 137
Crosstalk, 18, 129, 131
CRT (*see* Cathode-ray tube)
Current spike, 158
Current-voltage, (*see* i-V)
Curve tracer, 62

Depolarization angle, 143–145
Design:
  factors of, 13
  goals for connector, 19
  phases of, 13, 14
  worst-case, 21
Design-to-cost, 13
Design-to-power, 13
Design-to-reliability, 13
Design-to-size, 13
Detector, silicon, 37
Detector spectral response curve, 33
Dichroism, 139
Dielectric liquid, electrophoretic, 129
Diffuse source, 33
Diffuse surface, 43
Diffuser:
  marbled, 140
  MgO, 140
Diffusion, current, 103
Diffusion lengths in VLED, 105, 106
Diode:
  current injection efficiency, 104, 105
  efficacy of, 106
  plane, 132
  (*See also* Visible light-emitting diode)
Dipolar radiation, 139
Discharge, gas:
  ac, 59
  dc, 59

Display(s):
  character sizes of, 5, 7
  classification of, 1
  clock, 7, 9
  dot matrix, 1
  emissive, 1
  5 × 7 alphanumeric, 6
  5 × 7 VF, 127
  gas discharge, (*see* Gas discharge
    display)
  incandescent, 157–166
  multidigit, 7
  oven controller, 75
  passive, 1
  requirements, subjective for, 12
  segmental, 1
  7-segment gas discharge, 63
  size of, 5, 8, 13, 110
  specification for, 11
  stacking for, 7
  symbols for, 3
  temperature of, 22
  vibration of, 22
  VF, construction of, 127–129
  VLED: configurations for, 110–112
    sizes of, 110
Display bibliography, 1
Display comparison, 13, 15
Display contrast, 31
Display family, 2
Display mounting (*see* Mounting, display)
Display parameters, specifiable, 12
Display spectrum, 36
Distortion, VLED lens, 110
Dot matrices, 3, 6
Double-pole, double-throw (DPDT) switch,
    143, 144, 154
Double refracting structure, 137
DPDT switch (*see* Double-pole, double-
    throw switch)
Drive circuit(s):
  ac, gas discharge, 83, 87
  alphanumeric VLED, 120
  anode driver, 78
  constant-current, 80, 115, 117
  control diode, 80
  dc, 78
  low-impedance, 77
  pulse heated, 131
  segment driver, 78
  transistor, 79
  VF, 131
  voltage doubler, 79
Drive current, segment, 9

Driver, low-impedance, 27
Driving waveform, LCD, 143–155
Duty cycle, 148, 160
Dye lot samples, 45
Dynamic impedance, 61
Dynamic scattering, 137

ECD (*see* Electrochromeric display)
Efficacy, diode, 106
EL (*see* Electroluminescence, efficiency of)
Electric tuning eye, 123
Electric vector, 137
Electrical parameters:
  gas discharge, 77
  temperature dependence of, 77
Electrochromeric display (ECD), 2
Electroluminescence (EL), efficiency of, 104
Electrooptic light valve, 140
Electrophoretic cell, 127
  dielectric liquid, 129
Electrophoretic coating:
  anode, 124
  filament, 123
Electrophoretic display (EPID), 2
Emissivity:
  blackbody, 157
  VF filament, 124
End-of-life, LCD, 142
End-of-life criterion, 21
  gas discharge displays, 75
  LCD, 142
Environmental stress, 75
Environmental test, 20, 21
EPID (*see* Electrophoretic display)
Epitaxial fabrication step, 113
Epitaxy:
  liquid-phase, 106
  vapor-phase, 106
Equivalent circuit, gas discharge, 62
Esters, 137, 139
Excitation, gas, 59
Exhaust, VF, 129
Expansion joint seal, 142
Eye response, 33, 108, 111
Eye sensitivity to incandescent display, 157

Fabrication, VLED display, 113, 114
Failure(s):
  bathtub curve, 75
  catastrophic, 129
  field, 22

Failure mechanisms:
  gas discharge display, 74
  LCD, 141, 142
  VF, 129
  VLED, 114
Failure modes, 20, 21
Failure rate, incandescent display, 159
Faraday dark space, 59
Fate chart, charge carrier, 102
FFF (*see* Flicker fusion frequency)
Field failure, 22
Figure of merit, voltage, 146
Filament(s):
  aging of, 129
  anode geometry of, 124
  broken, 159
  incandescent, 157–166
  negligible evaporation of, 159
  pulse-heated, 131
  space charge limited, emission of, 125
  VF, 123
    stabilization of, 129
Filament temperature, incandescent display, 164
Filter(s), 18
  neutral-density, 43
  plastic, 45
  VLED, 108, 110
Filter display cover, 42
Filtering, incandescent display, 157
Flat-panel display, 68
Flicker, 74
  definition of, 55
  observer reactions to, 53
  slow display, 148
Flicker fusion frequency (FFF), definition of, 55
Fluorescent lamp:
  positive column in, 60
  spectrum of, 46
Fonts:
  by display type, 7
  for 7-segment displays, 7
Footlambert (fL) definition of, 55
Forward voltage, VLED, 116
Frame times, 148
Full select pulse, 152–155
Fusion frequency, critical, 16, 53, 55

Gallium arsenide (GaAs), band gap for, 100
Gallium arsenide phosphide (GaAsP):
  alloys of, 99–110
  band structure of, 100
  composition of, by color, 99

Gallium phosphide (GaP):
  band gap for, 100
  band structure of, 100
  as VLED material, 99
Gamma point, 100
Gas depletion, 67
Gas discharge:
  equivalent circuit, 62
  i-v relationship, 61
    sparking voltage, 61
  minimum current, 62
  parameter estimation of, 95–97
  scaling, 67
  segment driver, 78–81
  terminology for, 64
Gas discharge display, 59–97
  flat-panel, 1, 68
Gas lasers, 60
Gases:
  argon, 65
  helium-neon, 65–69
  krypton 85, 65
  liquid mercury vapor over, 65
  metastable ion in, 65
  neon, 59, 65
  neon-argon, 67
  nitrogen, 69
  noble, table, 65
  Penning mixture, 65, 69
  radioactive, 65
  tables for, 96–97
  tritium, 65
  xenon, 65
Getter, 73, 129, 158
Glare:
  definition of, 55
  surface, 43
  veiling, 43
Glare shield, 18
Glow discharge, 63
  color of, 68, 70
Glow discharge current density, estimation of, 97
Glow discharge tube, 59
Glow region, 59
  abnormal, 60
  normal, 60
Grid(s):
  cutoff voltages of, 126
  etched, 129
  potentials of, 125
  transparency, 124

Halation, definition of, 56
Half-select pulse, 152–155
Hemotropic structure, 138
Highlighting, 42
Human factors, definition of, 29

i-v:
  in plane diode, 132
  segment, incandescent, 164
i-v curve:
  symmetrical, 24
  VLED, 103
i-v relationship, gas discharge, 61
Illuminance, 30, 31
Incandescent display, 157–166
Incandescent display parameter
  estimation, 164–166
Incomplete cathode coverage, 61
Indium oxide anode, 72
Injection efficiency, diode, 104, 105
Inrush current, 166
Integrating sphere, 39, 42
Interelectrode space, gas discharge, 63
Interface, user-display, 29
Ion bombardment of phosphor, 129
Ion pumping, 67
Ionization potentials, 65

Junction:
  depth of, 106
  injection, 105
Junction current, 103

Keep-alive cells, 77
Krypton 85, 65

Lambertian surface, 31, 34
Laser-excited photo luminescence, 104
LCD (see Liquid crystal display)
Lead frame, 19
Leak checking, VF, 129
Leakage, 129
  segment-to-segment
Leakage current, 103
Legibility, 50
  green vs. red, VLED, 48
Legibility recommendations, 52
Legibility research, tools of, 51
Lens:
  bubble, 110
  cylindrical, 110
  viewing angle of, 110

Life test, 20
  accelerated, 20
  in-circuit, 21
  rolling, 21
Lifetime, VLED material, 100
Light, extraction from VLED, 106
Light bulb, 157
Light emission, neon-argon, 67
Light pipe, VLED package, 110
Light valve, 140
Line-of-sight electrons, 131
Line spectra, neon-argon, 67
Liquid crystal display (LCD), 137–155
  driving waveform, 143–155
  life of, 137
  materials for, 137
  multiplexing, 147–155
  no-selection, case of, 149
  power requirements of, 151
  response time, 143–146
  segment drive pulse, 143
  temperature coefficient, 146
  uniformity, 140
  viewing angle, 143
  viscosity, 146
Liquid phase epitaxy, 105
Lot acceptance, 20
Luminance, 31
  comparison of green vs. red, VLED,
    48
  definition of, 56
  incandescent segment, 158
Luminance build-up, definition of, 56
Luminance contrast ratio, definition of,
  56
Luminance conversion factors, 40
Luminosity (luminous efficacy), 35

Magnifier, 110, 112
Marbled diffuser, 140
Matrices:
  confusion, 51
  dot, 3, 6
MBBA (twisted nematic), 145
Mercury to retard sputtering, 60
Mercury-coating procedure, 74
Mercury spots, 74
Mesophase, 137
MgO, 140
Micropressor interfacing, VLED, 121
Minimum current, gas discharge, 62
Miscibility, alloy, 100
Modulation transfer function (MTF), 51
Mother board, 19
Mottled segment, 74

Mounting, display:
  adhesive for, 18
  procedures for, 18
  rigid mount for, 18
MTF (modulation transfer function), 51
Multidigit stacks, 7
Multiphase clock, 81
Multiplex operation, incandescent display,
  159, 160
Multiplexed displays, VLED, 117
Multiplexing, 16
  LCD, 147–155
  limits of, 24–27
Multiplexing voltages, optimum pulse set
  of, 152, 153

Nematic mesophase, 137
Neon-argon mixture, 60, 63
  spectrum of, 67
Neon gas, 59, 65
Neon glow, color of, 68, 70
Neon signs, 60
Nit, definition of, 56
Nitrogen, 69
  GaP, 105
  in VLED material, 101
No-selection, case of, LCD, 149
Nonradiative lifetime, 100
Non-sag tungsten, 164
Normal glow region, 60
Number of test units, 21

Observer-display interactions, 54
Off-state voltage, 143–155
On-state voltage, 143–155
Operating margin, 70
Operating temperature, incandescent
  display, 159
Optimum pulse set, LC, 147
Oven controller displays, 75
Overlay paste, VF, 129
Overvoltage, 75

Package, VLED:
  light pipe, 110
  multidigit, 110
  reflector, 110
Package cleaning, plastic, VLED, 115
Page-storage memory, CRT, 161
Parallel addressing, 14
Parameter estimation:
  gas discharge, 95–97
  incandescent display, 164–166

Parameters, gas discharge, variation with time, 75
Paschen's law, 65
    notation for (neon), 68
PBAB (twisted nematic), 145
PC board (*see* Printed-circuit board)
pd, 66
Penning mixture, 65
Persistence, definition of, 56, 57
Phases, design, 13, 14
Photoluminescence, 104
Photometer:
    calibration of, 37
    construction of, 37, 38
Photometric conversion factors, 40
Photometric measurements, 37
Photometric quantities, 37, 38
Photometry, 33
Photon generation rate, 102
Phosphor(s):
    aging of, 129
    brightness of, 125
    characteristics of, 161
    grain sizes of, 126
    ion bombardment of, 129
    quantum efficiency of, 126
    spectra of, 126
    VF, stabilization of, 129
Phosphor-coated anodes, VF, 123
Phosphor-coated cells, gas discharge, 68
Phosphor lag, 126
Planck's equation, 33, 34
Plane diode, 132
Point source, 33, 39, 42, 140
Polarization, 137, 139
Polarizer, 139–155
    circular, 43
Positive column, 59, 60
Power, incandescent display, 157
Power supplies well-regulated, 77
Precision glass, 142
Printed-circuit board, 18
    gas discharge, interconnection, 73
Processing steps, VLED, 113–115
Pulled-bond problem, 114
Pulse, LC, 147
        segment drive, 143
        select, 148–155

Qualification test, 21
Quantum efficiency:
    VF phosphor, 126
    VLED, 100, 117
Quick-disconnect sockets, 22

Radiance, definition of, 56
Radiation, thermal, 157
Radiative lifetime, 100
Radiative transition, VLED alloy, 101
Radio-frequency interference (rfi), 77
    shielding against, 78
    sweeping for, 77
Radiometric quantities, 39
Ratio, aspect (*see* Aspect ratio)
Recombination:
    band-to-band, 101
    current, 103
    surface, 104, 106
Recombination coefficient, 101
Reflection, 42
    cover, 43
    specular, 43
Reflector package, VLED, 110
Reliability:
    contact, 19
    VLED, 114
Reliability testing, 21
Repetition rate, incandescent display, 159
Resolution, 13
rfi (*see* Radio frequency interference)
rms (*see under* Root mean square)
Rolling life test, 21
Root mean square power, incandescent segment, 159
Root mean square voltage, definition of, 148

Scaling:
    current density, brightness, 9
    gas discharge, 67
Scanning electron microscope for cathodoluminescence excitation, 104
Schiff bases, 137, 139
Scratches, 43
Screen, CRT, division of, 160, 162
Secondary emission, 60
Segment(s):
    identification method for, 5
    shiny, 43
Segment drive pulse, LCD, 143
Segment driver, gas discharge, 78–81
Segment isolation, incandescent display, 159
Segment-to-segment leakage, 88–95
Select pulse, 148–155
Serial addressing, 14
Sharpness, 51
Shock, 22, 129
Shuttered cover, 43
Silicon, 99

Size:
    character, 51
    display, 5, 8, 13, 110
Slow-to-enter digits, 74
Smectic mesophase, 137
Sockets, quick-disconnect, 22
Soldering, temperature limits during, 115
Source:
    diffuse, 33
    point, 33, 39, 42
    standard, 33
Sparking voltage, gas discharge, 61
Specification(s):
    blemish, 13
    display, 11
    not in data sheet, 14
Spectral brightness, 140
Spectrum:
    display, 36
    neon-argon, 67
Specular reflection, 43
Spot welding, filament, 129
Sputtering, 65
Stabilization, VF filament and phosphor,
    129
Stacks, character, 7
Standard candle, 33
Standard detector, 33
Standard eye response, 33
Standard source, 33
Standards, cosmetic, 23
Starting, gas discharge, 77
State diagram, 154
Status flag, 7
Stefan-Boltzmann constant, 34, 158
Stress:
    current, 21
    environmental, 21, 75
Strobing (*see* Multiplexing)
Subjective brightness, definition of, 57
Subjective display requirements, 12
Surface recombination, 104, 106
Surround, 39, 42
Sustaining electrons, discharge, 60
Sweeping, rfi, 77
Symbol identification, rate of, 50
Symbols:
    display, 3
    special, 7, 8

Temperature:
    color, 30
    VLED brightness and, 114–117
Temperature coefficient, LC, 146

Temperature rise, 22
Test(s):
    cosmetic, 21
    environmental, 20, 21
    in-service, 21
    life (*see* Life test)
    qualification, 21
    typical display, 22
    typical times for, 23
Test methods, 19–23
    accelerated, 21
Test set, 20
Test units, number of, 21
Testing:
    diagnostics and, 22
    philosophy of, 19
    reliability, 21
Thermal expansion, differences in, 18
Thermal radiation, 157
Three-body interaction, 101
Threshold, 143–155
    temperature dependence of, 146, 147
    transmission, 148, 150
Threshold sharpness, 146
TI-150, 42
TI 3500, 67, 75
TIL 302, 119
TIL 303, 119
TIL 305, 6
Tiger striping, 74
Time(s):
    bounce and, 145
    digit, definition of, 64
    dwell, 148
    entry, 151
    fall, LCD, 145
    frame, 148
    ionization, 79
        definition of, 64
    new digit ionization, definition of, 64
    reionization, definition of, 64
    response, LCD, 143–146
    rise, LCD, 145
    segment blanking, definition of, 64
    segment heating and cooling, 165, 166
    segment off, 159
    turn-on or turn-off, incandescent display,
        157–166
Tin oxide anode, 72
tn (*see* Twisted nematic)
Transient peak for cold starting, 160
Transmission threshhold, 148, 150
Transparency, grid, 124
Transparent anode, 72
Triode, 123

Triode equivalent voltage and spacing, 133
Triode theory, 132
Triple carbonate, 129
Tripler, 143
Tritium, 65
Tungsten, 164
Turbid material, 137
Twisted nematic (tn), 137–155
  cross-aligned, 137
  MBBA, 145
  PBAB, 145
  positive anisotropy of, 145

Uniformity, LCD, 140

Vacuum fluorescent display (VF), 123–133
  geometry of, 123
  leak checking in, 129
  overlay paste, 129
  stabilization, of filament and phosphor, 129
Vapor-phase epitaxy, 106
Veiling glare, 43
VF (*see* Vacuum fluorescent display)
Vibration amplification, 22
Viewing angle:
  LCD, 143
  lens, 110
Viscosity, LC, 146
Visibility, 29, 30
Visible light-emitting diode (VLED), 99–122
  area of, 103
  common anode, 116
  common cathode, 116

Visible light-emitting diode (VLED) (*Cont.*):
  current degradation of, minimized, 103
  green vs. red, legibility of, 48
  i-v curve of, 103
  light extraction from, 106
  reliability, 114
  zinc-diffused, 113
Visible light-emitting diode display:
  materials for, 99
  micropressor interfacing, 121
  multiplexed, 117
  nonlinear light emission and, 103, 104
  operating principles of, 99–110
  packages, 110
    plastic, cleaning of, 115
  processing steps, 113–115
Visual acuity, 41
VLED (*see* Visible light-emitting diode)
Voltage:
  instantaneous, 148
  off-state and on-state, 143–155
  rms, definition of, 148
  wall, 69
Voltage figure of merit, 146
Voltage transfer curve, 69

Wall charging, 69
Wall voltage, 69
Waveform, driving, LCD, 143–155
Worst-case designs, 21

X-point, 100
Xenon, 65

Zinc-diffused VLED, 113